Functional anatomy in sports

Functional Anatomy in Sports

JÜRGEN WEINECK

Translated by

Thomas J. DeKornfeld, M.D.

YEAR BOOK MEDICAL PUBLISHERS, INC.

Chicago • London

This book is an authorized translation from the third German edition published and copyrighted by perimed Fachbuch-Verlagsgesellschaft mbH, Erlangen, Germany. Title of the German edition: *Sportanatomie*.

Library of Congress Cataloging-in-Publication Data
Weineck, Jürgen, 1941-
 Functional anatomy in sports.

 Translation of: Sportanatomie.
 Includes bibliographies and index.
 1. Sports—Physiological aspects.   2. Anatomy, Human.
I. Title.   [DNLM: 1. Exertion.   2. Movement.
3. Muscles—anatomy & histology.   4. Sports.
WE 103 W423s]
RC1235.W4413   1986   612'.044   86-5623
ISBN 0-8151-9193-6

Sponsoring Editor: Linda A. Miller
Manager, Copyediting Services: Frances M. Perveiler
Production Project Manager: Carol Ennis Coghlan
Proofroom Supervisor: Shirley E. Taylor

1 2 3 4 5 6 7 8 9 0 KC 90, 89, 88, 87, 86

Preface

Functional anatomy has been an integral part of the curriculum in physical education for many years, and is now becoming recognized as an essential subject in the education of athletic trainers. As a component of sports biology and athletic training theory, it is regarded as a significant contribution to the understanding of athletic motion studies and athletic career development.

Yet there is a notable shortage of publications to meet the need for a *functional anatomy textbook* in athletics. The textbooks of anatomy, used in medical education, are both too expensive and frequently couched in incomprehensible terms, totally unsuitable for the nonmedically trained athlete and for athletic training personnel. It is the purpose of this volume to examine the functional-anatomic basis of almost *all the sporting events of both summer and winter Olympics,* and to present this material in a format and language that makes the field of *sports anatomy* accessible to all those who have been frustrated by the complexities of technical jargon.

The *first chapter* of this volume lays the foundation for the understanding of the subsequent sections. Since every stimulus affects the cells, and since every functional response of the entire body is based on the response generated by the smallest functional units of the body, namely the cells and the tissues, the volume begins with the study of these fundamental structures. The higher structures are discussed in subsequent sections.

In the *second chapter,* a general overview of the organ systems is presented, with particular reference to the mechanics of passive and active motion. This section also contains a summary of essential anatomic principles.

The *third chapter* discusses the individual muscles that are the important components of active motion. The functional anatomy of the skeletal system and of the ligaments is presented only to the extent necessary for the understanding of muscular function. The 158 illustrations are designed to facilitate the understanding of the function of the individual muscles in a graphic and meaningful fashion.

The *fourth chapter* attempts to integrate the function of the muscles, discussed individually in the previous section, in the performance of simple trunk and extremity movements. This section represents a bridge between theory and athletic practice.

The *fifth chapter* presents an analysis of the complex locomotor processes of the different athletic events. A description of almost all of the Olympic sporting events will enable the layman, unfamiliar with anatomy, to gain a general view of the various muscle groups that are necessary for achieving success in each individual sport.

The *sixth* and last *chapter* offers the interested *nonspecialist* a brief introduction to training and practice exercises. This section also offers suggestions for static and dynamic exercises to strengthen the various muscle groups, which are essential for the performance of the Olympic sports. The exercises necessary for those sports that are not included in this book can be easily derived from the above.

The purpose of this book is, therefore, to present anatomic theory in a practice-oriented fashion and to adapt it to assist in the development of education, performance, and training.

JÜRGEN WEINECK

Contents

1 A Brief Study of Cells and Tissues

General Cell Biology (Cytology)

Since every stimulus acts primarily on the cell, and since all higher functions depend on the behavior of this smallest functional unit, it seems appropriate to begin this volume on sports anatomy with a description of the general form and function of the cell. This will lead into a discussion of the higher structures in the following sequence:

The *cell* → groups of cells = *tissue* → structural and functional grouping into *organs* and finally into *organ systems* as they develop into the active and passive kinetic systems of the body.

The Structure of the Cell

In its simplest terms, the cell consists of a cell body (cytoplasm), a nucleus, and different subcellular structures that are important in the function and maintenance of the cell. Only the most important ones will be discussed.

As shown in Figure 1–1, the cell is enclosed by a *cell membrane* (known as sarcolemma in muscle cells). The cell membrane is a complex, highly specialized biologic structure that has a selective permeability for electrolytes and organic substances and that also has the ability to associate with other cells. All the processes involving active transport (e.g., the sodium-potassium pump) are located in the cell membrane.

The *cytoplasm* (sarcoplasm in the muscle cell) is a fluid rich in electrolytes and pro-tein that is the site for anaerobic energy metabolism (glycolysis), gluconeogenesis, glycogenolysis, and fatty acid synthesis. (Glycogen is the intracellular storage form of glucose.) The cytoplasm also contains the various energy storage units, such as glycogen granules and fat droplets.

The *endoplasmic reticulum* (sarcoplasmic reticulum of the muscle cell) extends from the cell membrane throughout the entire cytoplasma and represents an intracellular transport system to which, in part, small globular structures, the *ribosomes* are attached. The endoplasmic reticulum and the ribosomes are the site of protein synthesis. In the muscle cell, the sarcoplasmic reticulum plays an important role in the transmission of stimuli from the surface to the contractile fibrils (see p. 19).

The *cell nucleus* contains genetic material and has the ability to reproduce in an identical fashion. The nucleus contains the template for protein synthesis, and is thus responsible for protein synthesis in conjunction with the above mentioned ribosomes. The two together permit the growth of muscle cells (hypertrophy) by synthesizing additional protein at the time of body growth or of athletic training. The *mitochondria* are the factories of the cell, since it is in these structures that the oxidative metabolism of the energy-rich substrates take place. The mitochondria contain the enzymes of the citric acid cycle and of the oxidative cascade. It is in them that oxidative phosphorylation and energy production takes place.

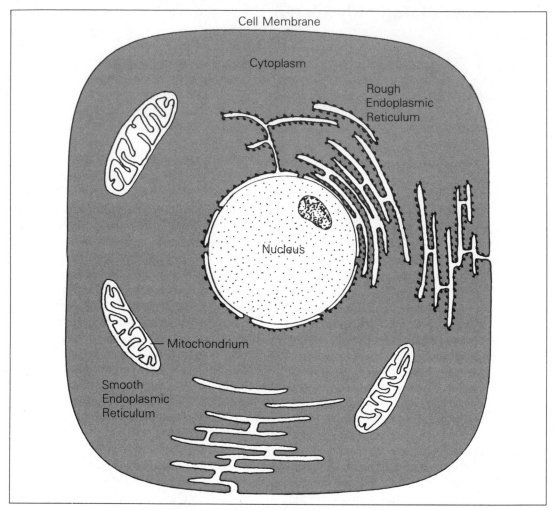

Fig 1–1.—Diagrammatic structure of a cell.

A Brief Tissue Biology (Histology)

A group of identical cells having the same function and differentiation are known as tissues. The tissue is a structural component of the total organism and performs certain specific functions.

There are four basic types of tissue:
Epithelial tissue
Connective tissue
Muscle tissue
Nervous tissue

All organs of the body *consist* of more than one tissue. Those cells that are responsible for the particular activity of the organ are known as *parenchymal cells.*

These parenchymal cells contrast with the *stroma cells* (interstitium), which serve to maintain the nutrition and external shape of the organs.

When increased specific demands are placed on an organ, the tissues can respond by hypertrophy (increase in the size of cells) and/or hyperplasia (increase in the number of the cells).

All tissues and organs originate from the three embryonal germ layers:

Epithelial tissue: ecto-, endo-, and mesoderm

Connective tissue: mesoderm

Muscle tissue: mostly mesoderm, some ectoderm

Nervous tissue: ectoderm

Epithelial Tissue

Epithelial tissue covers the external and internal surfaces of the body and serves as the most important functional component of the glands (glandular epithelium). In addition, epithelial tissue can also be adapted for special sensory functions (sensory epithelium).

Epithelial tissue: Surface epithelium
Glandular epithelium
Sensory epithelium

Surface Epithelium

Characteristics: Sheetlike layers of cells
Is located on the internal or external body surfaces
Lacks blood vessels
Is nourished by diffusion
Is separated from other tissues by a basal membrane

Functions: Protection (e.g., skin)
Secretion (e.g., renal epithelium)

Types: Flat, cuboidal, columnar, single layer, multiple layer, stratified, nonstratified, ciliated or nonciliated surface epithelium (Fig 1–2).

Glandular Epithelium

Characteristics: It serves to produce and to deliver certain substances (secretions and excretions). It consists of specialized epithelial cells.

Two types of secretory glands can be distinguished (*exocrine* or *endocrine*), depending on whether their product is used by the body or is eliminated from it.

1. Exocrine Glands

Their product (secreta) is delivered to the external or internal surface of the body, either directly, or through an excretory duct.

Examples: Sweat glands, salivary glands, tear glands, and digestive glands.

Depending on the *location* of the gland, the exocrine glands can be divided into *endoepithelial* or *exoepithelial* glands (located in the epithelium or in the subepithelial layers).

On the basis of the *process of secretion* there are the following glands. *Apocrine glands:* the secretion is collected at the tip of the cell and is then sloughed off, accompanied by some cytoplasm. (Example: lacteal glands.) *Holocrine glands:* The entire cell is sloughed off. (Example: the sebaceous glands of the skin.) *Merocrine glands:* the secretions are extruded in the form of droplets on the surface of the cell. (Example: sweat glands.)

On the basis of the *nature of the secretions* there are *serous* glands (thin, protein-rich secretion; example: the tear glands); *mucous* glands (thick, viscous secretions; example: the salivary glands at the base of the tongue); and *mixed glands* (example: salivary glands in the buccal area).

On the basis of the *structure* of the glands we distinguish *simple tubular* glands (example: small-bowel glands); *spiral glands* (example: sweat glands); *branching tubular* glands (example: gastric mucosal glands); *simple acinar* and *simple alveolar* glands (example: earwax glands and lung); and *compound* glands (example: salivary glands) (Fig 1–3).

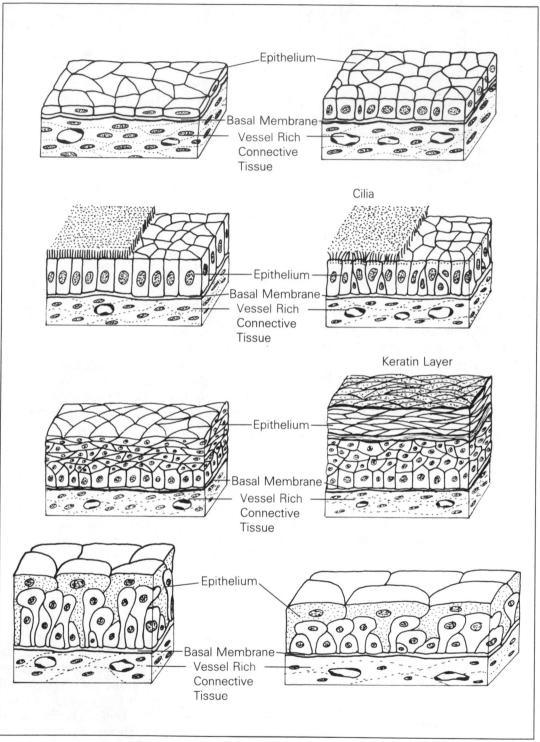

Fig 1–2.—The different types of surface epithelium: **1,** simple squamous epithelium, **2,** simple cuboid epithelium, **3,** simple columnar epithelium, **4,** stratified, ciliated epithelium, **5,** stratified, nonkeratinized squamous epithelium, **6,** stratified, keratinized squamous epithelium, **7,** transitional epithelium (not stretched) and **8,** transitional epithelium (stretched).

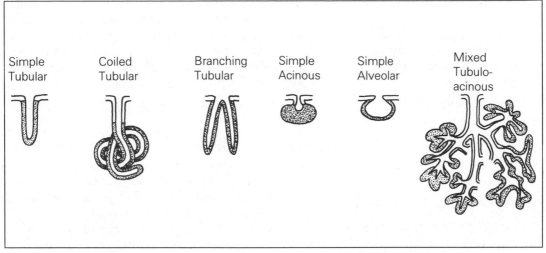

Simple Tubular   Coiled Tubular   Branching Tubular   Simple Acinous   Simple Alveolar   Mixed Tubulo-acinous

Fig 1–3.—Different types of glands (the secretory area of the gland is shown in dark outline).

Brief Digression on the Activity of the Sweat Glands

Sweat is the most dilute secretion produced by any gland and contains no more than 1% dissolved substance. This dilution is achieved by an active reabsorption of sodium in the distal end of the duct, from the isotonic solution produced in the proximal portion of the duct.

Heat adaptation, i.e., the result of physical conditioning, consists primarily in "teaching" the sweat glands to produce a larger amount of even more dilute sweat and thus to function more economically.

2. Endocrine Glands

These glands have no ductal system and deliver their output (secretions = hormones) directly into the blood stream (example: the pituitary gland, the thyroid, the adrenals).

Sensory Epithelium

The sensory epithelium consists of epithelial cells that have the ability to perceive and conduct specific stimuli. There are special sensory receptor cells for the sense of taste, smell, hearing, and sight.

Connective and Supporting Tissue

The connective and supporting tissues are distributed throughout the entire body and serve a variety of purposes. As bones, cartilages, and tendons, they form the "scaffolding" of the body. As connective tissues, they cover the organs and combine them into functional units. As stroma, they serve as the inner structure of the organs throughout which the functional parts of the organs, the parenchymal cells, are distributed. Certain special derivatives of connective tissue play a major role in immunological defense mechanisms. Finally, connective tissues also play an important role in storage (water, fat), transportation (between blood vessels and cells), and wound healing (scar formation).

The Divisions of Connective Tissue

The connective tissues can be divided into:

Mesenchyme (embryonal connective tissue)
Gelatinous tissue (jelly of Wharton) (embryonal connective tissue)
Reticular connective tissue
Fatty tissue
True connective tissue

Mesenchyme

Mesenchyme is present only in the fetus and is the basic substance from which other tissues evolve.

Gelatinous Connective Tissue

This substance is found only in the umbilical chord and in the dental pulp of children.

Reticular Connective Tissue

This substance receives its name from the fact that it forms a three-dimensional network (reticulum = network), which serves as the scaffolding of reticulolymphatic organs (spleen, lymph nodes, and bone marrow). In addition to their supporting function, the reticulum cells also serve as phagocytes (cells that ingest and digest cellular debris and foreign substances) and in the formation of free cells (see below).

Fatty Tissue

Fatty tissue makes up 10% to 20% of body weight and is distributed throughout the body. It performs a variety of duties:
Mechanical protection (e.g., as cushioning for the soles of the feet)
Closure of stomata
Shaping the human body
Insulation
Storage of energy and water

True Connective Tissue

Connective tissue consists of homogenous, amorphous *ground substance* and of an interconnected network of *fibers* that have varying structure and function. Both the ground substance and the fibers are produced by the *connective tissue cells* in different amounts and proportions, depending on functional requirements.

The Connective Tissue Cells

The connective tissue cells can be divided into fixed and free cells. The *fixed* connective tissue cells, also known as fibrocytes, produce the basic components of the connective tissue fibers and of the most amorphous intercellular substance.

The *free* connective tissue cells (e.g., white blood cells, histiocytes, plasma cells, and so on) do not participate in the formation of the intercellular substance and are components of the immunologic defense mechanism (RES, reticuloendothelial system). To some extent, they are ''satellite cells'' that settle in significant numbers in the tissue spaces and loops of connective tissue, where they are capable of *phagocytosis* (removal of foreign substances, e.g., bacteria).

The Intercellular Substances

The intercellular substances are composed of unformed and formed components. The unformed component consists of the *ground substance* and the formed component consists of the *fibers*.

1. The Ground Substance

The essential components of the ground substance are polysaccharides and proteins. The consistency of the ground substance is determined by the polymerization of these complexes. The ground substance serves as a matrix for fibers (see below) and, due to its viscosity, prevents the spread of foreign particles through the network.

In old age, there is a decrease in the ground substance. This leads to dehydration of the extracellular spaces and to decrease in tissue turgor. This, in turn, leads to the development of skin wrinkles.

2. The Fibers

The type of the connective tissue that appears throughout the body is determined largely by the type and structure of the fibers it contains. The fiber structure of the tissues depends on the mechanical forces and stresses that act on it. Any change in these stresses or forces leads to a modification and to a readaptation of the structure of the connective tissue fibers.

Three types of connective tissue fibers can be distinguished: collagen fibers, elastic fibers, and reticular fibers.

Collagen Fibers

Collagen fibers are present in practically all parts of the body and represent the largest fraction of all connective tissue fibers. They are the building blocks of the body, used particularly in all areas where traction forces are required. Because of this, they are very resistant to stretch and their resistance to traction is 6 kilopascals (kp) per square millimeter. The ultrastructure of the fiber in decreasing order is: fiber, fibril, microfibril (Fig 1–4).

Note: The diameter of the fiber and fibril varies considerably from tissue to tissue and depends on age and on the stresses to which they are exposed.

Repeated usage increases the diameter of the fiber, age decreases it.

Clinical implication: whenever a joint has to be immobilized for a longer period of time (e.g., following a fracture), the collagen fibers of the ligaments and capsule will shorten and this will lead to a temporary stiffening of the joint. Exercise will usually remedy this situation.

The Elastic Fibers

The elastic fibers are fundamentally different from the collagen fibers both in their structure and in their properties. They are significantly thinner; they branch and form a three-dimensional network. Their outstanding characteristic is their elasticity (up to 150% of their resting length).

The fibers return to their resting length once the traction (pull) has stopped. Since this allows a spontaneous contraction of organs rich in elastic fibers, it serves to reduce the work of the musculature and contributes to a general saving in effort and energy.

Elastic fibers are found principally in those organs that must change their shape regularly. Such organs are the lungs, the subcutaneous tissues, and the particularly elastic ligaments such as the ligamentum flavum of the spinal column.

Old age and inactivity leads to a decrease in the elasticity of the elastic fibers.

Reticulin Fibers

The reticulin fibers are the finest of all fibers in the human body. They are both elastic and reversibly expansible. Their

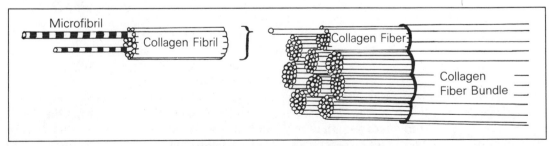

Fig 1–4.—The structure of a collagen fiber.

meshwork structure is affected by expansive and compressive forces. The reticulin fibers serve as the covering for numerous organs (e.g., liver, glandular, fat, and muscle cells) and as a scaffolding for others (e.g., bone marrow).

The Types of True Connective Tissue

1. Loose Connective Tissue

Loose connective tissue is the most widely distributed tissue in the human body, since it is distributed among and between all organs, blood vessels, and nerves, and since it serves as a filler and as connecting substance wherever necessary. Since it fills the interstitial spaces, it is frequently referred to as interstitial connective tissue. It is *characterized* by large intercellular spaces that contain *much amorphous ground substance* and *many free connective tissue cells*.

Loose connective tissue consists of a three-dimensional network composed of elastic fibers, reticulin fibers and, primarily, collagen fibers. This structure and composition allows both mobility between neighboring organs and tissues and also a reasonable amount of stability. Together they allow for movement as well as a return to the resting state.

The connective tissue cells, *fibrocytes,* are located singly, between the fibers.

Note: If one part of the body is injured, the fibrocytes can migrate from the loose connective tissue and cover the wound surface. They then form a scar, i.e., a connective tissue replacement for another type of tissue. The loose connective tissue has marked regenerative powers and is thus particularly suitable for wound healing.

2. Dense Connective Tissue

The dense connective tissue contains *little ground substance* and *few free connective tissue cells*. It is *rich in fibers* and relatively poor in cells. Compared with loose connective tissue, it has a low metabolic rate.

Since it is primarily composed of collagen fiber bundles, it is very resistant to mechanical stresses. For this reason, it is found primarily in those areas of the body that are particularly prone to stretch and to compression forces (example: ligaments, fascia, muscle, and tendon sheaths).

Supporting Tissues

The supporting tissues determine the shape of the body by virtue of their strength and consistency. They are divided into tendons, cartilage, and bone.

Tendons

The tendons serve to transmit muscular forces to the target organs, i.e., bones and joints. The ability to do this rests on the collagen fibers (tendon fibers) and on their specialized arrangement.

Advancing age leads to a 20% reduction in strength and in stretchability. A reduction in the number of tendon cells and in ground substance, combined with an increase in fibers and with a laying down of fat, is a characteristic feature of the aging process.

The thin, flat tendons are known as aponeuroses. The collagen fibers are arranged in parallel rows and are slightly wavy, while at rest. Between the collagen fibers are found the tendon cells (fibrocytes) arranged in rows, and shaped by the constraints of space into a shape that has earned them the designation of "winged cells" (Fig 1–5).

The structure of the tendon, in ascending order, is as follows: tendon fiber → primary bundle → secondary bundle → tendon. The primary bundles, the secondary bundles, and the tendon itself are surrounded by a connective tissue layer known as the internal and external peritendineal sheath (Fig 1–6). These sheaths contain nerves and blood vessels.

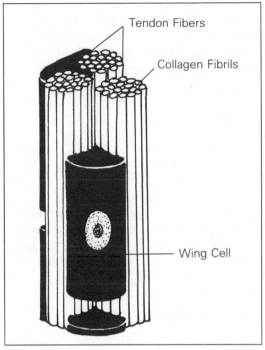

Fig 1–5.—Partial representation of a tendon.

The Attachment of the Tendon to Muscle and Bone

The Attachment of the Tendon to Muscle (Simplified Representation).

The tendon attaches the muscle to bone. The connection between muscle fibers and tendon fibers, also known as *musculotendineous transition,* is accomplished by a deep invagination of the tendon fibers into the muscle and by an attachment of the tendon fibers to the basal membrane of the muscle cells (Fig 1–7).

The Attachment of the Tendon to Bone

The attachment of tendon to bone, also known as *insertion,* represents the transition of the tendon to the osseous target organ. The function of the tendon, therefore, is to transmit the pull, generated by the contraction of the muscle, to the bone. The specific structure of the insertion area helps to ex-

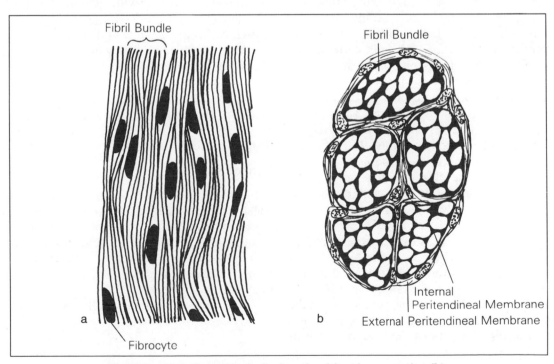

Fig 1–6.—A tendon in longitudinal section **(a)** and cross section **(b).**

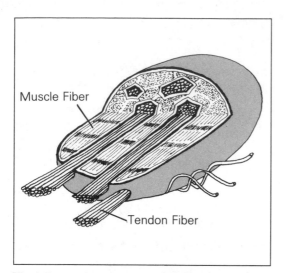

Muscle Fiber

Tendon Fiber

Fig 1–7.—Attachment of a tendon to a muscle fiber.

plain the intermediary role the tendon plays as a link between the active and passive components of the motor system. First of all, the tendon should never be torn loose from the muscle. To prevent this from happening, there is a rigid interdigitation between the collagen fibrils and the muscle fibers (see Fig 1–7). In addition, the resting fiber is slightly wavy. This, in combination with the elastic fibers present in the tendon, permits a gradual transmission of the pulling forces, which must first overcome this elastic resistance. This prevents and protects against the sudden impact of the stretch effort. On the other hand, this elasticity also creates problems at the tendon-bone interphase, since the elasticity constant—an indication of the ability of a substance to stretch—is quite variable (the order of magnitude is $\times 10$).

This problem is resolved by the introduction of a cartilaginous zone, as a buffer area for physical forces, between the tendon and the bone in the area of insertion. A gradual increase in cartilage cells toward the point of insertion permits a gradual adaptation of the elasticity of the tendon to that of bone.

From a mechanical point of view, the function of the insertion site is to create an equilibrium between two systems of differing elasticity (Becker and Krahl, 1978, Fig 1–8).

Cartilage

Similar to all other connective and supporting tissues, cartilage consists of cells *(chondrocytes)* and *intercellular material* (ground substance). There are three types of cartilage: fibrous cartilage, hyaline cartilage, and elastic cartilage (Fig 1–9). All three types of cartilage are characterized by the presence of *chondrones*. A chondrone is a well-defined area of cartilage, consisting of one or more chondrocytes surrounded by a capsule and an area high in mucous content, known as the lacuna.

Fibrous Cartilage

The intercellular substance of fibrous cartilage consists primarily of dense collagenous connective tissue with numerous parallel fibers and little amorphous ground substance. There are few chondrocytes.

Fibrous cartilage is very resistant to all kinds of forces—traction, pressure, and shear. When fibrous cartilage is under sustained pressure, it responds by a relative increase in the collagen fiber content (examples: annulus fibrosus of the intervertebral discs, menisci).

Hyaline Cartilage

In hyaline cartilage, the intercellular substance consists of numerous collagen fibers that are embedded in the ground substance and that are covered by it (the fibers are not visible under the microscope). The intercellular substance also contains numerous multicellular chondrones.

The ability of the joint cartilages to with-

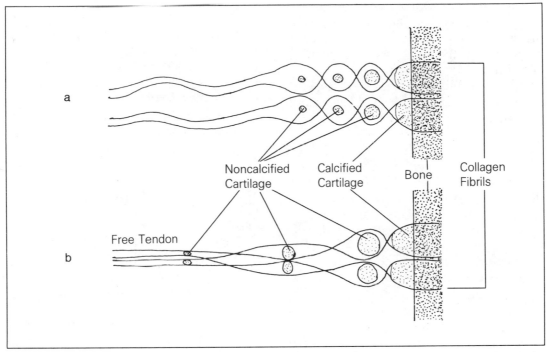

Fig 1–8.—Representation of the insertion zone of a tendon: at rest **(a)** and under tension **(b)** (after *Becker* and *Krahl*).

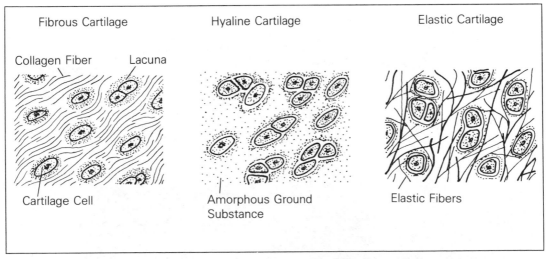

Fig 1–9.—Representation of the three types of cartilage (for details, see text).

stand pressure, traction and shear forces is due to the presence, and special structure, of the numerous chondrones and to the arrangement of the collagen fibrils they contain (Fig 1–10).

Note: Conditioning leads to a hypertrophy of the articular cartilage, which in turn, improves its elasticity.

Three millimeters is considered the critical thickness for cartilage from the point of view of being provided with metabolic sub strates (Franke, 1979). The fact that the cartilage of the patella may be as much as 6 mm in thickness explains the relative frequency of degenerative changes in this cartilage (patellar chondropathy), since this thickness is at the outer limits for the diffusion of the substances needed for normal metabolism.

The destruction of cartilage may be due to endogenous causes (e.g., individual reduction in weight-bearing ability) or to exogenous causes (e.g., excessive loading of the cartilage during training by performing deep knee bends while lifting weights).

Elastic Cartilage

The intercellular substance of elastic cartilage consists of collagen fibers and of a network of elastic fibers embedded in an amorphous ground substance. The elastic fibers make the flexibility of this type of cartilage possible (example: the cartilage of the external ear).

Common Properties of the Three Types of Cartilage

Cartilage contains neither blood vessels nor nerves. Nutrition is provided by diffusion from capillaries on the surface of the cartilage or, in joint cartilage, from the synovial fluid.

The aging process leads to a decrease in the water content of the cartilage, which leads to a decrease in elasticity.

Cartilage is a tissue of low metabolism (bradytrophic tissue) and limited powers of regeneration.

The activity of chondrocytes is affected by hormones. Testosterone increases it and cortisone decreases it.

If the chondrocytes are stimulated by increased load requirements, they respond by

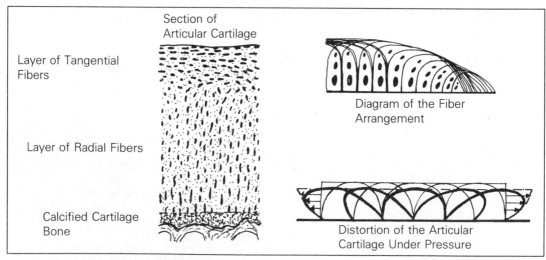

Fig 1–10.—Organization and function of the collagen fibers in articular cartilage (after *Benninghoff* and *Goerttler*).

increasing their metabolism and adapt by increasing the synthesis of collagen and ground substance.

Bone

The bone is the most stable tissue in the human body. Its resistance to pressure is ten times greater than that of cartilage. This is due to the incorporation of inorganic components (85% calcium phosphate, 10% calcium carbonate, and 5% magnesium and alkaline salts) into an organic ground substance. The organic components of bone consist of 95% collagen fibers and 5% amorphous ground substance. In addition, bone contains cells, the so-called *osteocytes,* which are responsible for the production of bone, for the destruction of bone, and for producing structural changes in bone. Histologically, there are two types of bone: *lamellar* and *reticular* bone. Since the latter is present in the adult only in a very few places (e.g., in a part of the temporal bone) only lamellar bone will be discussed in this section. The long bones of the extremities are typical examples of lamellar bone.

The Structure of a Long Bone

The long bone consists of two major parts: the diaphysis and the epiphysis. The *diaphysis* is the shaft of the long bone, i.e., a hollow tubular structure composed of hard bony tissue (substantia compacta), which encloses the bone marrow in the cavity of the tube.

The *epiphyses* are the two articulating ends of the long bone, which are covered with hyalin cartilage. The interior of the epiphysis consists of a spongy scaffolding of bone spicules (substantia spongiosa).

In still growing bone, there is a tissue layer between the diaphysis and epiphysis consisting of hyalin cartilage and known as *epiphyseal symphysis.* It is at this point that the longitudinal growth of bone takes place until the growing period is completed and ossification of this symphysis takes place (see page 17).

The lateral layers of the bone are externally the periosteal membrane, the bone proper, and internally, the bone marrow.

The Periosteum

The periosteum covers the bone on all sides, with the exception of the articulating surfaces and the insertion of ligaments and tendons. It consists of an inner reproductive layer (stratum osteogeneticum) and an outer, fibrous layer (stratum fibrosum). The collagen fibers of the periosteum penetrate the bone *(Sharpey's fiber)* and fix the periosteum to the bone.

The periosteum is well supplied with blood vessels and nerves and serves the following functions:

- Nutrition of the bone through blood vessels.
- Protection of bone, since it surrounds the bone with a firm, yet elastic membrane. Because of its rich nerve supply, it warns of mechanical impact on the bone (pain). Example: periostitis due to unusual or excessive stress.
- Regeneration: By virtue of its bone-producing cells, it participates in the regeneration of bony tissue and in the formation of new bone (callus) following a fracture.

Bone Tissue

The external osseous material consists of a more or less solid layer, the *substantia compacta,* which varies in consistency depending on the functional stresses to which it is exposed. The internal osseous material is a spongy structure, *substantia spongiosa,* consisting of a network of bony spicules.

The Substantia Compacta (Bony Cortex)

The external osseous tissue is composed of general lamellae (inner and outer base lamellae), of osteones that have special

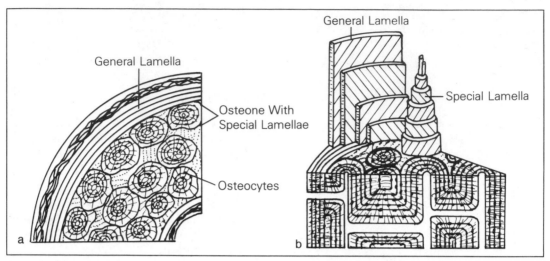

Fig 1–11.—**(a)** Cross section of a lamellar bone, **(b)** spacial representation of the general lamellae and special lamellae.

lamellae, and connecting lamellae (i.e., the lamellae between the osteones) (Fig 1–11).

The structural unit of lamellar bone is the *lamella* (3 to 7 μ in thickness). In these lamellae, the collagen fibers are parallel to each other. The direction of the fibers changes from lamella to lamella, usually at an angle of 90°. The osteocytes are located primarily on the edges of the lamellae and have long extensions, which lie in fine bone canaliculi and which are freely interconnected with each other (Fig 1–12).

The General Lamellae

The general lamellae encompass the bone in its entirety and form several layers, both on the exterior and interior of the bone. The exterior general lamellae are adjacent to the periosteum and the interior lamellae are oriented toward the marrow cavity.

The Osteone

The *osteone* consists of a central canal (the Haversian canal) surrounded by concentric lamellae, the so-called *special lamellae*. The central canal contains blood vessels from which bone receives its nu-

trients by diffusion. In addition to the longitudinal Haversian canals, there are also transverse canals, the so-called Volkmann's canals.

The Connecting Lamellae

The connecting lamellae are really lamellar fragments that fill in the spaces between the osteones in the substantia compacta of the diaphysis of the long bones. The struc-

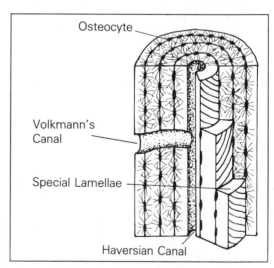

Fig 1–12.—Structure of an osteone.

ture of the connecting lamellae corresponds to the special lamellae.

The Substantia Spongiosa

The substantia spongiosa provides the internal structure of the bone and consists of a spongy aggregation of bony spicules that are derived from lamellar fragments. The spongy material is generally oriented along the main lines of stress and serves as a container for the red marrow (Fig 1–13).

Bone Marrow

There are two types of bone marrow: *yellow* and *red*. *Yellow* marrow is 96% fat and, hence, is also referred to as fatty bone marrow. It is located in the longitudinal marrow cavity of the long bones and serves to occupy the space in this location.

Red bone marrow is found in the interstices of the substantia spongiosa and is the most important hematopoietic organ of the body.

Bone Development

Bone development (ossification) starts in the mesenchyme. There are two types of bone development: the *direct,* or *membraneous,* and the *indirect* or *cartilaginous.* In the latter type, the first step is the development of a cartilaginous skeleton, which is later replaced by bone.

Membraneous Ossification

The following steps take place:
- Individual mesenchymal cells develop into osteoblasts, i.e., bone-generating cells.
- Each osteoblast secretes a ground substance (osteoid) around itself.
- The osteoid incorporates collagen fibers, which are developed outside the cells. The precipitation of calcium salts leads to the formation of ossification centers.
- Through further osteoid formation and

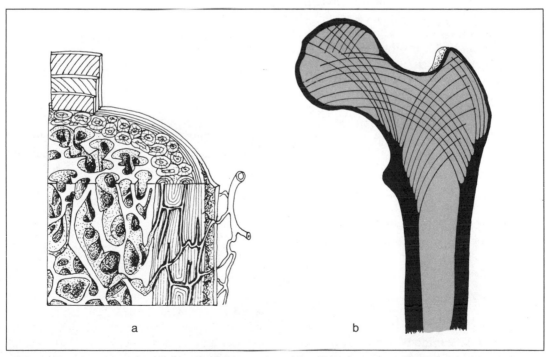

a   b

Fig 1–13.—Representation of the substantia spongiosa (**a**) and the stress-dependent arrangement of the bony trabecula (**b**).

calcification, the ossification centers develop into bone spicules, which in turn, gather together into spongy bone tissue.

- Finally, the external and internal bony layer develop (example, membranous ossification: certain bones of the skull).

Note: Membraneous ossification occurs only in the fetus. To allow growth and development of the infant and child, the bones must undergo repeated developmental changes. This is accomplished by the interaction of osteoclasts (bone-destroying cells) and osteoblasts (bone-generating cells).

Cartilaginous Ossification

Most of the human bones are developed through cartilaginous ossification. The precursor of cartilaginous ossification is a model made of hyaline cartilage. The change into bone is accomplished by *pericartilaginous* and *endocartilaginous ossification*. In the *pericartilaginous* ossification of long bones, ossification takes place in an annular form, around the cartilaginous model. The end result is a bony sleeve in the diaphyseal area that can further grow in diameter by laying down bony tissue in an appositional fashion. The metamorphosis of the cartilaginous model itself is accomplished by *endocartilaginous* ossification (supplemental bone development). In this process, the cartilage cells are destroyed by chondroclasts (cartilage-destroying cells) and then replaced, first by a bony lattice, and later, by lamellar bone through the action of osteoblasts. Both chondroclasts and osteoblasts evolve from mesenchymal cells, which penetrate the cartilage at the surface of the bony sleeve, along the blood vessels in the dense connective tissue.

Bone Growth

Longitudinal Growth

The pericartilaginous bony sleeve acts as a splint around the cartilage and prevents lateral growth while exerting pressure that forces the cartilage toward the two open ends of the bony envelope. This leads to the endocartilaginous, longitudinal growth of bone.

Note: With time, the bony spicules in the area of the future marrow cavity are destroyed by osteoclasts. This takes place along the entire length of the bone up to the transitional zone between diaphysis and the two epiphyses, which are present in children and adolescents, and which are known as the growth plates (epiphyseal plates). In this way, the marrow cavity is developed.

As long as the growth plate is in existence, longitudinal endocartilaginous growth can take place. Once the plate is ossified, no further growth can take place.

Transverse Growth

Transverse growth (thickening) occurs exclusively through pericartilaginous ossification. It originates in the periosteum and is referred to as appositional growth.

Muscle Tissue

The characteristic feature of the muscles is their ability to contract. The muscle is an organ of motion that is composed of a large number of muscle cells. The muscle cells are characterized by the contractile protein structures, the myofibrils. Every muscle tissue is accompanied by connective tissue, which is responsible for the interaction of muscle cells, and which transmits the contraction of the muscle cells (muscle fibers) to their environment.

On morphologic and structural grounds, two types of muscle are identified. These are the striated muscles and smooth muscles. The striated muscles are further divided into skeletal muscle and cardiac muscle. The most obvious difference between striated and smooth muscle is the presence of striations in striated muscle (hence the name), which can be seen by both light microscopy and electron microscopy. Smooth muscle lacks these striations. Furthermore, striated muscle is primarily innervated by the somatic nervous system and the smooth muscle by the autonomic nervous system. The contractions of striated muscle (i.e., voluntary contractions) are rapid and straight, while the contractions of smooth muscle are slow and wormlike.

Cardiac muscle is in a special class since, even though it is striated, it is innervated by the autonomic nervous system and has an inherent automaticity. Every muscle tissue consists of elongated muscle cells having one or more nuclei, surrounded by cytoplasm.

The striated muscle cells are arranged in parallel rows, and their nuclei are located along the lateral wall of the cell. The cardiac muscle cells are branched and are attached to each other via intercalated disks!

Their nuclei are located centrally in the cells. The nuclei are also centrally located in the smooth muscle, which are the smallest of the muscle cells. While cardiac muscle cells have a diameter of 15 μ and striated muscle cells have a diameter of 50 μ, the smooth muscle cells have a diameter of only 5 μ (Fig 1–14).

The Smooth Muscle

Smooth muscle is found in those locations where the requirements are for a slow and sustained contraction, rather than for a rapid contraction. For this reason, smooth muscle is found primarily in blood vessels, gut, and bladder.

Striated Muscles

1. Skeletal Muscle

The contractile element of the skeletal muscle consists of muscle fibers. These muscle fibers are organized into primary bundles, secondary bundles, and, finally, into the muscle proper, by cover membranes containing collagen and elastic fibers.

Skeletal muscle fibers can attain a length of 15 cm.

The basic constituent of the muscle fiber is the striated muscle fibril, which consists of thin (actin) and thick (myosin) filaments.

Between the myofibrils there is a mitochondrium-poor sarcoplasm, and a *well-developed, smooth sarcoplasmic reticulum* (the so-called L-system) in which the calcium ions, necessary to trigger a contraction, are stored. When a nerve impulse is delivered by the T-system (Fig 1–15), the calcium ions penetrate the sarcoplasm and cause the myofibrils to contract.

There is only a small amount of granular (rough) sarcoplasm and some ribosomes that are responsible for protein synthesis. This fact explains why mature muscle has only very limited regenerative ability and why destroyed muscle fibers are replaced primarily by connective tissue scars. This happens, for example, when muscle fibers are torn.

One of the peculiarities of the muscle cell and muscle fiber is the fact that there are different muscle fibers, all having specific and different functions. Omitting consideration of an intermediate fiber, there are two primary muscle fiber categories:

- The white (light), thick, and "fast" fiber, also known as FT (fast-twitch fiber). Its main role is in forceful and intensive muscular activity.
- The red, thin, so-called "slow" ST fibers (slow-twitch). This muscle fiber is used for low intensity muscular activity.

Example: The *gastrocnemius* muscle, used primarily for rapid, vigorous activity (e.g., jumping) consists mainly of FT fibers, while the *soleus* muscle, used primarily for sustained activity, is composed mostly of ST fibers. In consequence of their different functional requirements, the different muscle fiber types also differ in their

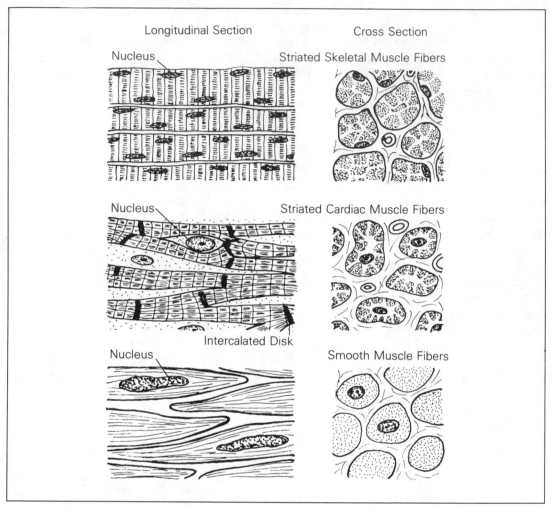

Longitudinal Section   Cross Section

Nucleus   Striated Skeletal Muscle Fibers

Nucleus   Striated Cardiac Muscle Fibers

Intercalated Disk

Nucleus   Smooth Muscle Fibers

Fig 1–14.—Longitudinal and cross section of striated skeletal muscle, striated cardiac muscle, and smooth muscle.

metabolic properties. The FT fibers are rich in high energy phosphates and glycogen, and are, accordingly, provided with enzymes of anaerobic energy production. The ST fibers are also rich in glycogen, but are provided generously with the enzymes required for aerobic metabolism.

The presence and relative distribution of the different types of muscle fibers is determined genetically. In the "born" sprinter, the FT fibers are preeminent, while in the "born" long-distance performer, the ST fibers are dominant.

The Contractile Process

A contraction takes place when the actin filaments are drawn up between the myosin filaments. This telescopic displacement of the filaments is accomplished by a coupling of the tips of the myosin fibrils and the thin action filaments, triggered by the influx of calcium ions into the sarcoplasm. A sweeping motion of the tips of the myosin fibrils (very much like the motion of oars in rowing) pulls the thin action filaments up, between the thick myosin filaments, resulting

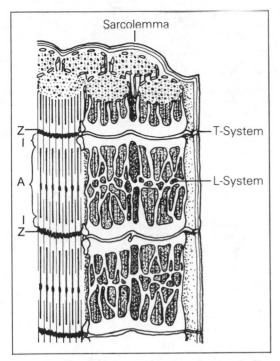

Sarcolemma

Z
I
A
I
Z

T-System

L-System

Fig 1–15.—Partial representation of a muscle fiber consisting of several myofibrils. The tubules of the transverse *(T)* system are directly connected to the longitudinal *(L)* system of the sarcoplasmic reticulum (see text). On the left, representation of a sarcomere (section between two Z stria) after removal of the sarcoplasmic reticulum.

in a visible shortening of the entire muscle (Fig 1–16).

Note: The musculature is constantly in a certain tension state. This "tone" of the musculature assures, on one hand, the erect posture of humans (the extensor muscles of the back are always just sufficiently contracted to keep the spinal column straight), and, on the other hand, also assures the continuous readiness for immediate muscular activity.

2. Cardiac Muscle

The cardiac musculature (see Fig 1–14) has the following characteristics:
- The cardiac muscle cells are striated, very much like skeletal muscle, but they are irregularly branched and only about 100-μ long.
- The cardiac muscle fibers are connected end-to-end through the intercalated discs.
- The nucleus of the cardiac muscle fiber is located centrally.
- There is a mitochondrium-rich sarcoplasm between the cardiac muscle fibers. Thirty percent of the volume of cardiac muscle cells consist of mitochondria (skeletal muscle has only 5%), which makes the aerobic energy storage and availability possible for this "sustained action" performer.
- The excitation of cardiac muscle is initiated by a specialized muscle tissue, the excitatory musculature, which has spontaneous activity. In addition, the cardiac musculature is innervated by the autonomic nervous system.

Nervous Tissue

Among the fundamental properties of living organisms is the ability to be stimulated, to conduct a stimulus, and to respond to stimulation. In unicellular organisms, all three of these properties are found in a single cell. In metazoan animals and humans, stimulation is the responsibility of a special conductive tissue, the nervous system.

Nerve tissue is composed of nerve cells, nerve fibers, and neuroglia.

The Nerve Cell

The nerve cells, also known as ganglion cells, are found in the grey matter of the brain (approximately 150 billion cells), in the spinal cord, in the spinal ganglia, and in the ganglia of the autonomic nervous system. They occur in different forms (unipolar, bipolar, pseudounipolar, and, most frequently, multipolar) and are of different sizes (4 to 120 μ).

An aggregation of nerve cells is called a *ganglion* in the periphery and a *nucleus* in the CNS.

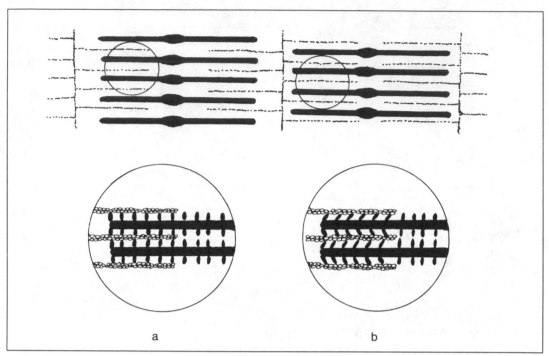

Fig 1–16.—Simplified representation of a contraction. Thick filaments = myosin, thin filaments = actin, **(a)** relaxed state, **(b)** contracted state.

The nerve cells conduct and integrate impulses. Each nerve cell consists of a cell body (pericaryon), which contains the nucleus and of extensions known as axons and dendrites.

The *pericaryon,* composed of the nucleus and surrounding cytoplasm, is the trophic center of the nerve cell. Its surface can respond to both positive and inhibitory stimuli. The cytoplasm contains the *Nissl bodies,* which are responsible for the synthesis of structural and transport proteins (Fig 1–17).

Since the nerve cells have lost the ability to divide (reproduce), destroyed nerve cells cannot be replaced by new nerve cells.

The Nerve Fiber

The nerve fibers are extensions originating in the nerve cells. There are two types: axons and dendrites.

The Axon

Each cell has one axon that serves to conduct an impulse from the cell to the periphery. Depending on the size of the organism, the axon can be as long as 1 m (e.g., the axons in the sciatic nerve reach from the sole of the foot to the spinal cord).

The Dendrite

The other extensions of the nerve cell, the dendrites, are shorter, branched, and varying in number. They conduct impulses toward the nerve cell.

The smallest *structural unit* of the nervous system is known as the *neuron.*

The neuron is composed of the nerve cell, the axon, and the dendrites. This organization is also a trophic unit, since the axon and dendrites depend on their nerve cell for nourishment.

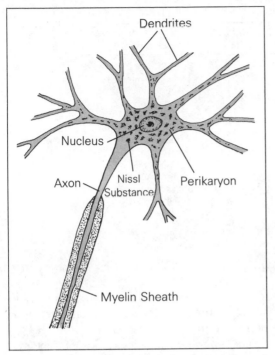

Fig 1–17.—Diagrammatic representation of a multipolar nerve cell.

The smallest *functional unit* of the nervous system is the *reflex arc*. In its simplest form, it consists of an afferent neuron that carries impulses from the periphery to the CNS, where the impulse is processed through a synapse, to an efferent, i.e., motor neuron. It is in such a way that an impulse is transmitted to a target organ, e.g., muscle.

Structure of the Nerve Fibers

Axons and dendrites have essentially identical structures (Fig 1–18). The axis cylinder lies in the middle and is surrounded by an external sheath, the neurilemma sheath, which serves to insulate the axon electrically. Depending on the structure of sheath, we distinguish myelinated (medullated) or unmedullated nerve fibers. The lipoprotein complex making up the medullary layer is also known as myelin.

The nerve fibers can be divided into groups on the basis of size, diameter, and conduction velocity (Table 1–1).

Structure of a Nerve Bundle

Most nerve fibers are running in bundles (Fig 1–19). In the CNS, these are called

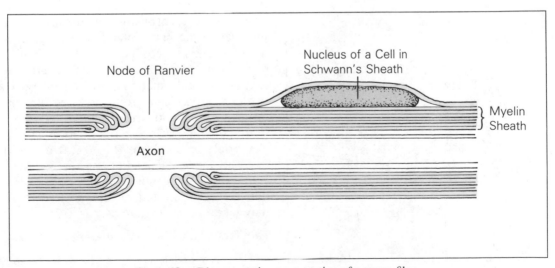

Fig 1–18.—Diagrammatic representation of a nerve fiber.

TABLE 1–1.—THE CLASSIFICATION OF NERVE FIBERS

GROUP	NERVE FIBER CROSS SECTION, μ	CONDUCTION VELOCITY, m/sec	EXAMPLE
Myelinated Fibers			
A	10–20	60–120	Efferent to striated muscle
A	7–15	40–90	Afferent from the skin (touch)
A	4–8	30–45	Efferent to smooth muscle
A	3–5	5–25	Afferent from the skin (temperature)
B	1–3	3–15	Preganglionic autonomic fibers
Nonmyelinated Fibers			
C	0.3–1	3–15	Postganglionic autonomic fibers

fasciculi; in the peripheral nervous system they are simply called *nerves.*

The nerves connect the peripheral parts of the body to the CNS. There are afferent nerves, i.e., those that carry impulses toward the CNS (sensory) and efferent, i.e., those that carry impulses from the CNS (motor). Most nerves are mixed nerves.

In the nerves, the nerve fibers are surrounded by connective tissue structures, which connect the bundles to each other and to their environment. These connective tissue structures are divided into endoneurium, perineurium, and epineurium.

The *endoneurium* is a loose connective tissue that surrounds the nerve fibers directly and that carries the nutrient capillaries and lymphatics.

The *perineurium* is a tough connective tissue sheath that surrounds groups of nerve fibers and protects them against stretch. The *epineurium* also protects against stretch and ties the numerous nerve fiber bundles into a single nerve. Sensory and motor fibers run together in the nerve bundles. If a nerve is transected (e.g., through trauma) one of two events can take place. The distal part can *degenerate* and function is lost, or the proximal axon part can grow, reunite with the distal portion, and accomplish a *regeneration* and reinnervation. Since the axon can grow only at a rate of 0.5 to 3.0 mm per day and since, depending on the site and extent of the injury the gap may be substantial, regeneration and regaining of function may take longer than a year.

The Transmission of Impulses

The transmission of impulses from the nerve to the target organ, e.g., muscle takes

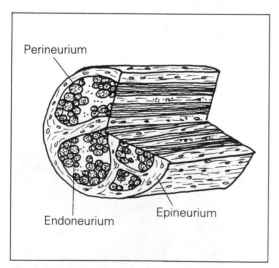

Fig 1–19.—Representation of a nerve fiber bundle.

place across specific contact areas that are known as *synapses* (Fig 1–20). The transmission of impulses is unidirectional. A synapse consists of three components:

- The terminal part of the axon, which ends in a slight swelling, and the synaptic knob, which contains the synaptic vesicles. These, in turn, contain the transmitter substance, which is required for the transmission of impulses.
- The synaptic cleft, across which the transmitter substance has to carry the impulse to:
- The postsynaptic membrane, which is part of the target organ and where the propagation of the impulse is generated by the transmitter substance.

Neuroglia

Neuroglia, or glia, is the connective tissue support structure of the CNS, in which the ganglion cells and nerve fibers are embedded. The astrocytes are the principal glial cells of the white and grey matter in the CNS, the oligodendrocytes are the principal glial cells of the peripheral nerves.

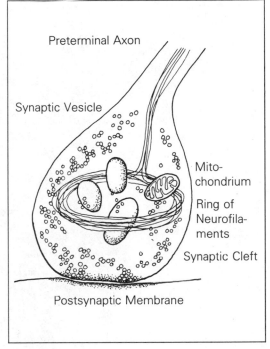

Fig 1–20.—Electron microscopic representation of a synapse.

Scene from the German Water
Skiing Championship.

2 The Passive and Active Locomotor System

Anatomic Nomenclature (Fig 2–1)

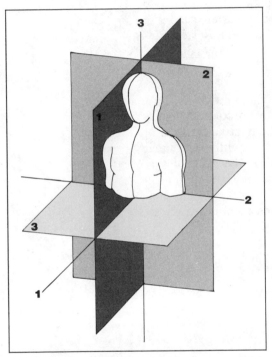

Fig 2–1.—The anatomic planes and axes of the human body. *1*, sagittal axis–sagittal plane; *2*, transverse axis–frontal plane; *3*, vertical axis–horizontal plane.

General Overview of the Locomotor System

The locomotor system consists of two components: the skeletal system and the muscle system. The skeletal system includes the bones, joints, and ligaments; the muscle system includes the muscles and their accessory components such as tendons, tendon sheaths, bursa, etc.

The ability of muscles to contract makes it possible to change the relative position of bones that are connected with each other through a movable joint. The skeletal system provides the passive motion component of the body, while the muscle system constitutes the active motion component.

The Passive Locomotor System

General Osteology and Arthrology

The Function of Bone

In addition to their protective function (brain, bone marrow), the bones serve two additional functions. They serve as a scaffolding that supports and sustains the soft-tissue organs, and they also serve as rigid levers for muscular activity.

The Types of Bone

Corresponding to their specific functions and to the demands placed on them, there are a variety of bone types. Some are tubular and long, like the bones of the extremities; some are flat and wide like the scapula, the pelvic bones, and the bones of the skull; some are short and cuboid like the vertebrae and the bones of the wrist and ankle.

The Adaptation of Bone to the Stresses and Strains of Athletic Activity

The development, growth, and maintenance of bone is affected through hormonal and mechanical regulatory mechanisms.

Mechanical stress, as for instance, through athletic training, constitutes a stimulus that can specifically alter the composition and general pattern of bone. There may be an increased mineralization of bone, a thickening of the cortical layer of the bone, and a strengthening of the trabecular structure of bone. This latter change in the trabecular system involves both a change in the structure of the trabeculae and, also, in their reorientation along the pressure-tension lines of the mechanical stresses (see Fig 1–13).

Mechanical stress leads not only to adaptive changes in the infrastructure of bones, but also to changes in their external configuration. The tuberosities, cristae, and tubercula, which serve for the insertion of muscles and tendons, can become larger, and indeed, the entire bone may change its shape under the influence of muscle activity. The tibia may serve as an example for the latter occurrence: this bone is primarily tubular at birth, but becomes triangular with increasing mechanical stress occasioned by assuming the upright position and walking. In children who are paralyzed from birth on, this change does not take place.

Note: Contrary to its effect on skeletal muscle, continued, unphysiologically high stress does not lead to further hypertrophy of bone, but to atrophy and demineralization, which may lead to *stress fracture*.

The Bones as Components of Joints

The Classification of Joints

The bones are connected with each other, either immovably or movably. On this basis, we can distinguish between synarthroses and diarthroses (immovable and movable joints).

The Synarthroses

A synarthrosis is a firm connection between bones with a connecting substance that allows no motion at all, or only a very limited one.

We can distinguish between the following:

1. Syndesmosis (ligamentous). Example: the firm connective tissue connection between tibia and fibula.

2. Synchondrosis (cartilaginous). Example: the cartilaginous connection of the symphisis pubis.

3. Synostosis (osseous). Example: the bony connection of the components of the sacrum.

The Diarthroses

A diarthrosis is a movable connection between bones across an articular cavity. The bone ends are covered with cartilage, and can move relative to each other and thus form a true joint.

The Structure of a Diarthrosis

The Articulating Surface

The articulating bone ends are covered with hyalin cartilage, which provides a smooth surface and minimizes friction. The articulating surface is either convex (we speak here of the *head of a bone*) or concave—this is known as an *acetabulum*. The acetabulum can be deepened by a cartilaginous lip around the acetabulum (glenoid labrum): examples are the shoulder and the hip joints. When the articulating surfaces are uneven, this lack of fit can be improved by cartilagenous discs, or menisci, placed into the joint cavity.

The Articular Capsule

The articular capsule is a connective tissue enclosure for the joint that provides an airtight cover for the joint cavity. It is attached to both articulating bones, usually at the edge of the articulating cartilaginous surfaces. The capsule consists of an inner and outer layer. The inner layer is smooth and secretes the synovial fluid that lubricates the articulating surfaces. The outer layer consists of collagenous connective tissue.

The Joint Cavity

In reality there is no such thing as a joint cavity, since the space between the articulating surfaces is, at best, a slit with the thickness of a hair. This is due to pressure (weight) on the joint, and to the pull of the muscles acting on the joint.

Classification of the Diarthroses

The mobility of a joint is determined by the shape of the articulating surfaces and by the arrangement of the articular ligaments (Fig 2–2).

Single Axis Joints

The *hinge joints:* allows flexion and extension, e.g., the joints between the segments of the fingers.

The *pivot joint:* allows rotation of the radial head in the proximal radioulnar joint, where the radial head sits in a concave area of the ulna (incisura radialis ulnae) and is held in place by an annular ligament. The motion permitted by this joint is pronation and supination of the hand.

Biaxial Joints

The *condyloid joint:* dorsiflexion and palmarflexion, and radial and ulnar abduction at the wrist.

The *saddle joint:* just as in a riding saddle, both articulating surfaces are concavoconvex. The single example for this type of joint is between the first metacarpal bone of the thumb and the trapezoid bone of the wrist.

The Triaxial Joints

The *ball joint:* the area of the acetabulum is smaller than the surface of the ball. The

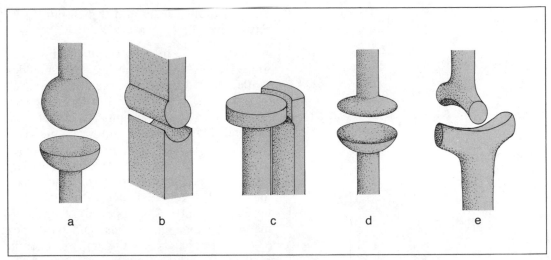

Fig 2–2.—Diagrammatic representation of the different types of articulation: (**a**) ball and socket joint, (**b**) hinge joint, (**c**) the pivot joint, (**d**) condyloid joint, (**e**) saddle joint.

example for this joint is the shoulder, which is both the most movable and most vulnerable joint in the body.

The *"nut joint."* In this joint, the acetabulum is deep enough to encompass the ball beyond its equator. In this way the hip joint is much less likely to suffer dislocation than the shoulder joint. (Translator's note: The American literature refers to both these joints as *ball and socket joints*.)

The Limiting Factors of Joint Mobility

Osseous Limitations

Motion is limited by the articulating bones themselves. In the elbow, for example, extension is stopped by the olecranon.

Ligamentous Limitations

The limits of motion are determined by the ligaments surrounding the joint. The ileofemoral ligament allows only very limited extension of the trunk when the femur is held rigid.

Muscular Limitations

Where muscles extend over several joints, motion will be limited in some joints

when the muscles are maximally contracted. Its impossible to make a fist when the wrist is maximally flexed.

General Overview of the Skeletal Structure of Humans

The human skeleton consist of the spinal column, the skull, the shoulder girdle, the upper extremity, the pelvic girdle, and the lower extremity (Fig 2–3).

The Spinal Column

The spinal column represents the central skeletal axis of the human body. It protects the spinal cord, holds up the head and permits its mobility, supports the shoulder girdle, and is the link to the pelvic girdle. It consists of 33 to 34 bony segments, the vertebrae, which are linked above the sacrum by the small intervertebral joints, the intervertebral disks, and a strong system of ligaments.

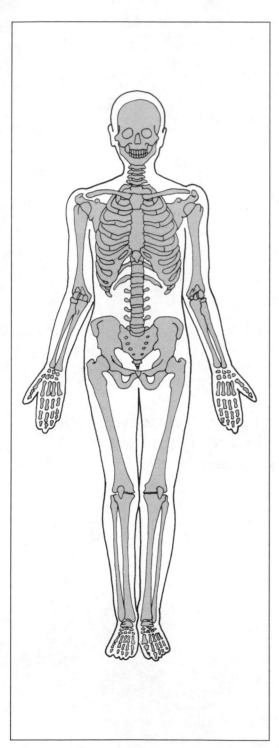

Fig 2–3.—Diagrammatic representation of the human skeleton.

The Shoulder Girdle and the Upper Extremity

The upper extremity is attached to the trunk by the shoulder girdle. The shoulder girdle, consisting of the scapula, clavicle, and sternum, is connected to the trunk in a peculiarly mobile fashion. It actually is suspended by a muscle sling and its only osseous connection with the chest is functionally a ball and socket joint. This arrangement is essential to permit the maximal mobility to the upper extremity. The upper extremity is composed of the humerus, the ulna, the radius, and the bones of the wrist, hand, and fingers. Its function is to serve as a tool for grasping, holding, and expression.

The Pelvic Girdle and the Lower Extremity

Similarly to the upper extremity, the lower extremity is also attached by a girdle to the skeleton of the trunk. The pelvic girdle consists of the two hip bones and the sacrum.

The lower extremity, which is connected to the pelvic girdle, consists of the femur, the tibia, the fibula, and the bones of the ankle, foot, and toes.

The significance of the pelvic girdle as a support structure is manifest in the size and strength of its components and in the fact that these components, originally separate, have fused together to form a single unit.

Tae-kwon-do done perfectly.
Attack and defense.

The Active Locomotor System

General Myology

The skeletal musculature is composed of approximately 400 individual muscles, widely differing in both form and size.

The Forms and Types of Muscle

A muscle can have one or more individual heads of origin that join together into a single tendon. We can distinguish between the following:
- Muscles with one head, e.g., *the brachialis m.*
- Muscles with two heads, e.g., *the biceps brachii m.*
- Muscles with three heads, e.g., *the triceps brachii m.*
- Muscles with four heads, e.g., *the quadriceps femoris m.*

A muscle may also have several bellies in one line, connected by tendons. Example: the *rectus abdominis m.*

A muscle may participate in more or less complex motions, depending on the number of joints that are covered by one tendon. We speak of uniarticular muscles, e.g., the *brachialis m.;* biarticular muscles, e.g., the *sartorius m.,* and multiarticular muscles, e.g., the *flexor digitorum profundus m.*

Depending on the *arrangement of the fibers* (Fig 2–4) we can distinguish:
- The *fusiform muscle,* e.g., the *biceps brachii m.*
  Note: The muscle belly tapers at both ends into the terminal tendons. The fibers that lie parallel to each other on the surface become pennate toward the interior.

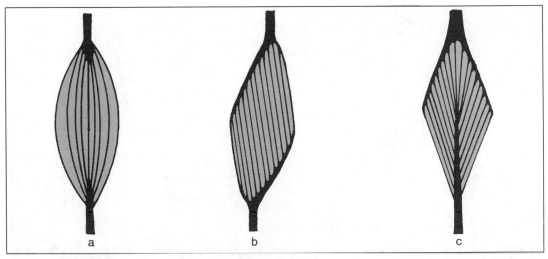

Fig 2–4.—The different patterns of muscle fiber organization: **a,** parallel fibers, **b,** simple feathering, **c,** double feathering.

- *Unipennate muscle*. Example: *extensor digitorum longus m.*
- *Bipennate muscle*. Example: *the quadriceps femoris m.*

It can also happen that the same muscle shows a different arrangement of fibers in its various portions. Thus, for example, the fibers of the *deltoid m.* are parallel to each other in the anterior and posterior portion of the muscle, while the middle portion boasts three to five bipennate tendons.

Muscle Mechanics

Vertical Lift and the Arrangement of Fibers

The *vertical lift* of a muscle is proportional to the length of the muscle fiber bundles and to the changes in their angle of insertion. The maximal contraction of a muscle is 50% of its original resting length. Additional shortening is made impossible by the mechanical glide mechanism of the muscle filaments. The muscles that have *parallel fibers* have a particularly long vertical lift, and are therefore designated as *speed muscles*. Example: the *biceps brachii m. Bipennate muscles*, with an obtuse angle

of insertion of the muscle fibers, are typically muscles with a short vertical lift, but with the ability to generate *maximal force*. These muscles are particularly important in *grasping* and sustained effort work. Examples: the *vastus medialis, lateralis, and intermedius muscles* of the *quadriceps femoris m.*

The strength of a muscle depends on the sum of the cross sections of its fibers and on their angle of insertion.

Anatomic and Physiologic Cross Section

One must distinguish between an *anatomic* and a *physiologic* cross section. The *anatomic* cross section is the cross section obtained by cutting the muscle at right angles to the longitudinal axis of the muscle. The *physiologic* cross section is the sum total of the cross sections of all the fibers in the muscle (Fig 2–5).

In muscles having all their fibers in parallel alignment, the anatomic and physiologic cross sections are identical. In all other muscles, the anatomic cross section is smaller than the physiologic cross section.

The maximal force that can be generated

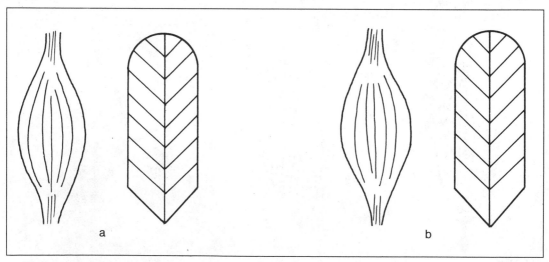

Fig 2–5.—Representation of the anatomic (**a**) and physiologic (**b**) cross section in a muscle with parallel fibers, where both cross sections are identical and in a muscle with double-feathered fibers, where the physiologic cross section is considerably larger than the anatomic one.

by a muscle is roughly equal to 6 kp/sq cm of the cross-sectional area. This cross-sectional, area-dependent force is also affected by such factors as age, sex, muscular coordination, motivation, etc.

Mechanical Self Regulation

Since the muscle becomes thicker during contraction, there has to be a special mechanism that allows this to happen without impairing the contraction process.

This mechanism is the so-called mechanical self-regulation (Benninghoff and Goerttler, 1975). It is the acute angle insertion of the muscle fibers onto the tendon that allows, through a change in this angle toward the less acute, the creation of enough space to permit the enlargement of the fibers during a contraction (Fig 2–6).

Origin-Insertion-Center of Rotation

The sites of attachment of the muscle are usually referred to as *origin*—fixed point, and *insertion*—mobile point. The effect of a muscle contraction can be derived from the knowledge of the origin and insertion. It is important to realize, however, that knowing the *origin* and *insertion* gives only a general idea of the effect of the muscle contraction and should not be considered as a rigid and inflexible fact in the practical assessment of motion.

In general, a certain point of the locomotor system is designated as the origin, if it is *immovable* in relation to a fixed point such as, for instance, the trunk. The insertion is considered to be the *moving part,* by virtue of having less mass.

In a number of motions, however, both fixed points of the muscle change position, and there may even be a reversal of the usual mechanism whereby the insertion of the muscle becomes fixed and it is the ori-

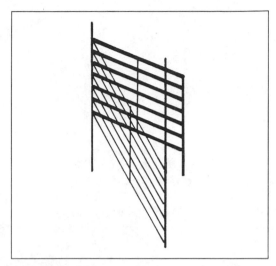

Fig 2–6.—Diagrammatic representation of the mechanical autoregulation of the muscle (after *Benninghoff* and *Goerttler*).

gin of the muscle that moves on contraction. This happens, for instance, during a front support mount on the horizontal bar, when the arms are fixed by their position on the bar and a contraction of the muscles causes the trunk to move toward the arms.

In judging the locomotor function of a muscle and of its different components, it is important to consider the relationship of the muscle to the *axis of rotation* of the joint it operates. If the muscle, or its tendon, lies *anterior* to the axis of rotation of the joint, it serves as an extensor, e.g., the *quadriceps femoris m.* at the knee. If it lies posterior to the axis, its effect is flexion.

It can also happen that during a contraction, the different parts of a muscle assume an antagonistic role. A classic example is the *deltoid m.* Below the axis of rotation, the anterior and posterior part adduct the arm (Fig 2–7,a); above the axis they abduct and elevate the arm and support the function of the middle portion (Fig 2–7,b).

Synergists and Antagonists

In every motion, a group of muscles functions either simultaneously or in

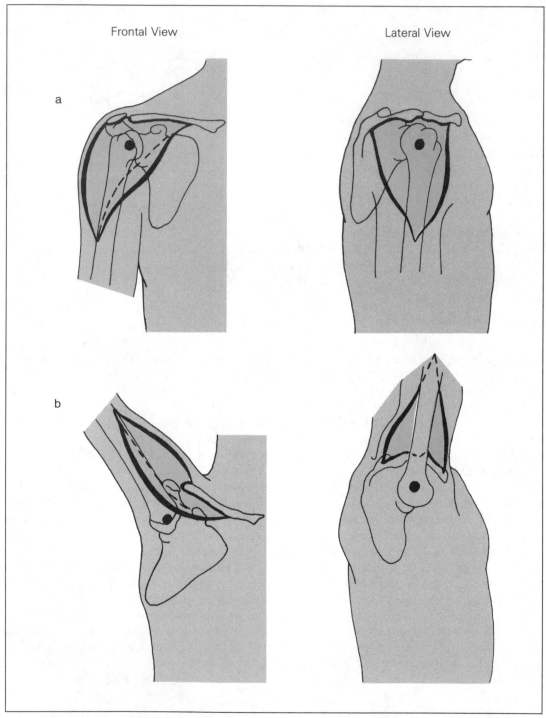

Fig 2–7.—Change in locomotor function of a muscle within a normal range of motion depending on its relationship to the rotational axis. Example: the deltoid m. **(a)** The muscle is located primarily under the rotational axis, **(b)**, it is above the rotational axis.

succession. Only rarely does a muscle contract all by itself. Muscles that cooperate in the accomplishment of a movement are known as *synergists*. Muscles that during the performance of a motion act in opposition to the general direction of the move are known as *antagonists,* even if the action is limited to a passive stretch. In effect, all motion is influenced by the interaction of *synergists* and *antagonists*.

Note: A muscle that is prestretched by its antagonist prior to a contraction can develop an increased contraction, as shown by the "wind-up" prior to a throw. If, on the other hand, a muscle is put into a position where only limited shortening is possible, the ensuing force of contraction will be severely limited.

The Accessory Structures of the Musculature and of the Tendons

Most muscles are connected to their osseous points of origin and insertion by *tendons*. Some muscles, however, do not originate from bone and do not insert into bone, but do so on the tough connective tissue sheets, the so-called interosseous membranes, which are a connective tissue extension of the bony skeleton.

The Tendons

Just as in muscle, there is a wide variation in the type of tendons, dependent on their function. There are long narrow tendons, as well as short and wide ones. Some tendons are flat and sheetlike (aponeuroses).

The Insertion of Tendons

The tendon is usually anchored to bone at a specially structured site, which as a rule is markedly calcified and which contains cartilaginous tissue (see p. 11).

Depending on the form of insertion and on the pull exerted by the muscle, the bone will develop a variety of protuberances. These can be circumscribed protuberances, i.e., tuberosities, tubercula, spinae, processes, and trochanters, or fine and coarse ridges, i.e., lines and cristae.

Protective Mechanism for the Tendons

In those areas where there is a significant mechanical load, we frequently find special structures such as sesamoid bones, bursae, or tendon sheaths. The *sesamoid* bones are ossified segments of the tendon, which strengthen the tendon and improve the traction mechanics of the muscle. The largest sesamoid bone in humans is the patella.

The bursae are small, fluid-containing sacks, which are found in all locations where muscle or tendon has to glide across bony eminences. They function as a bolster, which absorbs the tension on the tendon and protects it from the underlying bone.

Tendon sheaths have the duty of assuring that the tendons can glide smoothly in those areas, where this may otherwise be difficult. Such areas are those where tendons run for extended distances over bone, or where two or more tendons cross. They consist of tough connective tissue, and are provided on the inner surface with a secondary layer, which produces a synovial fluid. This fluid allows the motion of the tendons to be as free of friction as possible.

The Fasciae

The fasciae are tough, connective tissue sheets that surround individual muscles or muscle groups, and serve as a conduit for them. They ensure that the muscle or muscle group is in its proper position and thus ready for action. Occasionally, the fascia may serve as a site of origin or insertion for other muscles.

3 Individual Discussion of the Most Important Articulations

The Trunk

The trunk has two main functions: the protection of the various organ systems it surrounds and to serve as the base for the motion of the extremities and the support of the head. To maintain the erect posture of the body, the spinal column is provided with a dynamic equilibrium by the musculature of the belly and back (Fig 3–1).

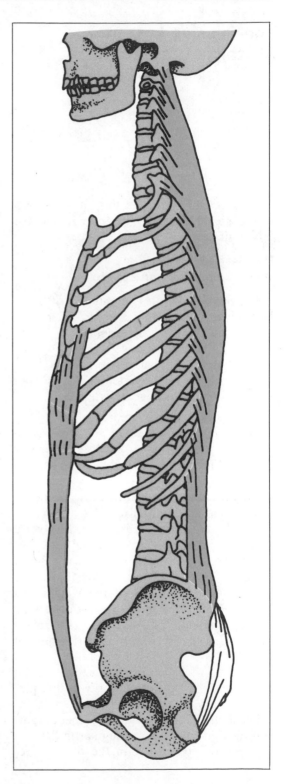

Fig 3–1.—The muscles of the trunk as a system of braces for the maintenance of the erect posture.

The Passive Locomotor System of the Trunk

The skeletal system of the trunk consists of the spinal column, the thorax, and the pelvic girdle.

The Osseous Structure of the Spinal Column

The spinal column is composed of 33 to 34 bony segments, the vertebrae. There are:

7	cervical vertebrae
12	thoracic vertebrae
5	lumbar vertebrae
5	sacral vertebrae
4–5	coccygeal vertebrae
33–34	total vertebrae

The five sacral vertebrae and the four to five coccygeal vertebrae are fused into single units known, respectively, as the sacrum and the coccyx.

All vertebrae are constructed according to an identical scheme and consist of a vertebral body, the vertebral arch, two transverse processes, a spinous process, and articulating surfaces. They differ from each other, however, depending on their location in the spinal column. The increasing weight that the lower vertebrae have to carry contribute to the structural changes that take place from the cervical to the sacral vertebrae (Fig 3–2).

Only the first two cervical vertebrae, the atlas and axis, do not conform to this pattern (see p. 71).

The *body* of the vertebra is the weight-bearing element. It is rectangular in the cervical region, triangular in the thoracic re-

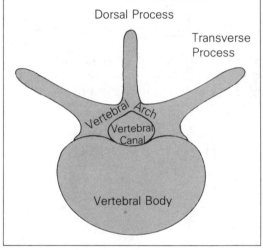

Fig 3–2.—Diagrammatic representation of the structure and form of a vertebra (superior aspect).

gion, and bean shaped in the lumbar area (Fig 3–3).

Between the 24 vertebral bodies above the sacral area there is, in each instance, an intervertebral disk, which functions as a shock absorber. The intervertebral disk consists of the nucleus pulposus that is surrounded by a fibrous band, the annulus fibrosus. The *annulus fibrosus* constitutes the major portion of the intervertebral disk. It consists of concentric and spiral cartilaginous fibers and connective tissue, which are anchored to the cartilaginous plate of the two neighboring vertebrae. This provides a particularly solid connection between the vertebral bodies. The *nucleus pulposus*

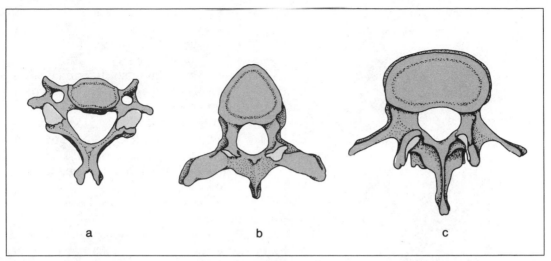

Fig 3–3.—Shape of the vertebral surface in the different parts of the spinal column: **(a)** cervical area, **(b)** thoracic area, **(c)** lumbar area.

serves to distribute the stresses between the vertebrae during flexion and extension of the spinal column. When bending forward, the nucleus pulposus is displaced posteriorly; when bending backwards, it shifts in an anterior direction, and similar action takes place on lateral bending of the spinal column. In other words, the intervertebral disk connects the neighboring vertebrae, serves as a shock absorber for the spinal column, and contributes to its mobility.

Ventrally and dorsally, strong longitudinal ligaments are attached to the vertebral bodies and help to maintain the natural S-shaped curve of the spinal column.

Digression: Degeneration of the Intervertebral Disk

Excessive athletic or other stresses or tearing (shear) forces lead to a decrease in height of the intervertebral disk through an erosive and *degenerative* process. This, in turn, leads to a relaxation of the longitudinal ligaments and to a loosening of the moving segments. The ensuing shift of the vertebral bodies leads, frequently, to a narrowing of the vertebral foramina and to a compression and irritation of the roots of the spinal nerves. This is the cause of the common, and highly variable, painful condition of "low back pain" (Fig 3–4).

The importance of the intact intervertebral disk on the weight-bearing potential of the spinal column can be illustrated by the following example: assuming that a person holds a 10-kg weight with an outstretched arm (lever arm length, 75 cm), the spinal column must distribute a weight of 150 kg over the extensor muscles in the lumbosacral area, since these muscles have a lever arm length of only 5 cm and the ratio of the weight arm to the lever arm of effort is 15:1. This entire weight rests on the intervertebral disks, in addition to the regular weight of the body, and they must both accept it and distribute it. If the intervertebral disks are destroyed, this entire force will act directly on the vertebral bodies.

The *vertebral* arch is attached to the *vertebral bodies* dorsally and forms the *vertebral canal* in which the spinal cord is located. The vertebral arch is the base for several processes.

• Two laterally directed *transverse pro-*

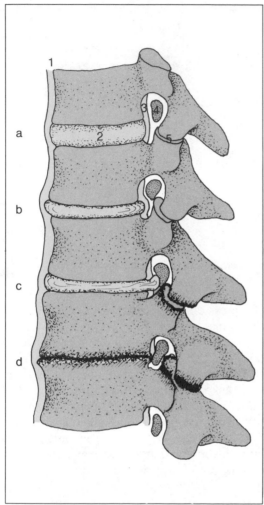

Fig 3–4.—Diagrammatic representation of the events during the degeneration of the intervertebral disk. (**1**, anterior longitudinal ligament, **2**, intervertebral disk, **3**, posterior longitudinal ligament, **4**, nerve root in the intervertebral foramen, **5**, intervertebral articulation.) **a**, Normal relationships in the intervertebral area, **b**, change in the relationship between vertebrae with loss of distance, loosening of the ligaments and distortion of the intervertebral foramen (exit part for the nerve root!), increased pressure on the intervertebral articulating surfaces, and anterior displacement of the intervertebral disk in the direction of the nerve roots, **c**, extrusion of the nucleus pulposus backward with pressure on the nerve, **d**, degeneration of the intervertebral disk with approximation of the neighboring vertebral bodies, formation of prongs along the edges, and protrusions on the bodies of the vertebrae and on the intervertebral articulations.

cesses. They support the ribs in the thoracic area, fuse with the rudimentary ribs in the lumbar area to form the costal processes, and form with the costal processes transverse foramina in the cervical area. These foramina contain major blood vessels.

- Two pairs of *articular processes*. It is through the corresponding upper and lower articular processes that the articulation of neighboring vertebrae takes place.
- One posteriorly directed *spinous process*.

The varying orientation of the articular processes and of the spinous process in the different segments of the spinal column determine to a large extent the mobility of spinal column (see Fig 3–33 p. 51).

The *transverse processes* and the *spinous process* serve as origin and insertion sites for muscles and ligaments. The pronounced development of the spinous processes in the lumbar spine indicate the leverage required in this area to maintain the erect posture of the spinal column.

It can be stated that the anterior portion of the vertebrae (the vertebral bodies) serve as a support column; the central portion (the vertebral arches) serve as protection for the spinal cord, and the posterior portion (the processes) serves as levers to assure the mobility of the spinal column.

The first two cervical vertebrae, the atlas and the axis, have a quite different structure, which requires special consideration (see p. 71).

The Shape of the Spinal Column

The human spinal column is not straight, but has characteristic curvatures in the individual portions of the column, namely the cervical lordosis (anterior curve), the thoracic kyphosis (posterior curve), the pelvic lordosis and the sacral kyphosis. These curvatures are related to the maintenance of the erect posture. In addition, the cervical lordosis contributes to the mobility of the head, while the lumbar lordosis contributes to the springiness of the thorax (Fig 3–5).

Lateral curvatures are pathologic and are referred to as scolioses.

The Mobility of the Spinal Column

The mobility of the spinal column is determined by the small vertebral articulations, which are structured differently in the cervical, thoracic, and lumbar regions (Fig 3–6).

The mobility of the individual vertebral joints is limited, but the summation of these limited movements provides considerable mobility to the vertebral column as a whole.

The Structure of the Vertebral Articulations

In the *cervical spine,* the articulating surfaces are flat and tilted slightly anteriorly. Since the joint capsules are quite loose, and since the orientation of the spinous processes is such that it does not interfere with mobility, there is the possibility of considerable flexion, extension, rotation, and lateral motion.

For this reason, the cervical spine is the most mobile part of the spinal column.

In the *thoracic spine,* the articulating surfaces are set obliquely and stand at a slight angle toward each other. This permits considerable rotation, but this motion is limited by the tight ligaments and by the ribs. Nevertheless, lateral motion, flexion, and extension are still possible. Extensive motion in the thoracic spine is sharply reduced by the long, overlapping, and downward-angled spinous processes.

In the *lumbar area,* the articulating surfaces are almost vertical and the articular processes face each other. This fact makes rotation in this area impossible, and even lateral motion is severely curtailed. This limitation in mobility serves to assure the posture and the ability to walk in the upright position. Extension in this area is excellent

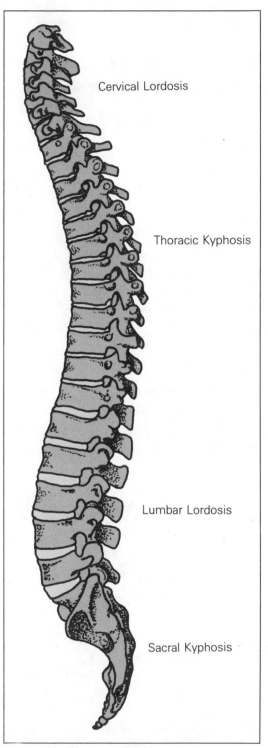

Cervical Lordosis

Thoracic Kyphosis

Lumbar Lordosis

Sacral Kyphosis

Fig 3–5.—Structure and form of the spinal column.

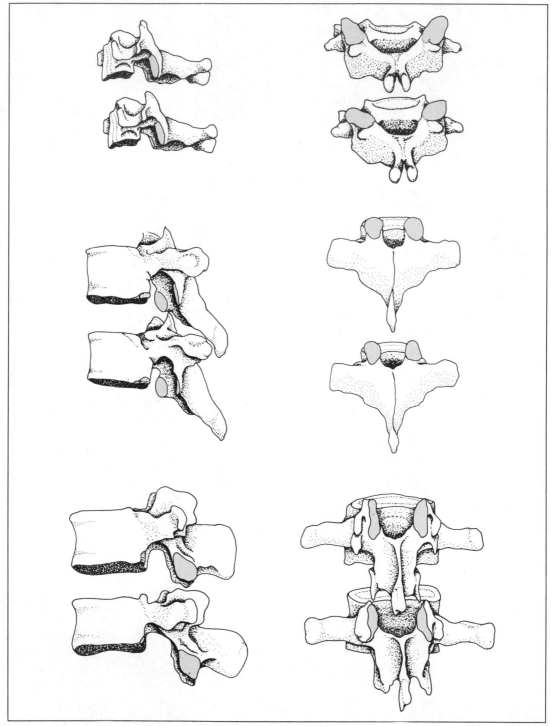

Fig 3–6.—Structure and location of the small intervertebral articulations in the lateral and dorsal view in the cervical region *(above)*, thoracic region *(center)* and lumbar region *(below)*. The articulating surfaces are indicated by shading.

European gymnastic
championship. Elena Muchina
on the balance beam.

(e.g., in strutting) and flexion is good. Extension and flexion are facilitated by the horizontal orientation of the spinous processes, which also provide good leverage for the powerful muscles of the lumbar spine region.

In summary, it can be said that the mobility of the spinal column decreases from top to bottom because of the increasing static load and the consequently increased need for stability.

The Ligamentous System of the Spinal Column

Tight ligaments substantially limit the mobility of the spinal column. The *longitudinal ligaments* are attached anteriorly and posteriorly to the vertebral bodies. The vertebral arches are connected by the *ligamentum flavum*. The bases of the spinous processes are connected by the *interspinal ligaments,* and the tips of the spinous processes are connected by the *supraspinal ligaments*. Finally, the transverse processes are connected by the *intertransverse ligaments* (Fig 3–7).

The spinal column is attached to the sacrum (a component of the pelvic girdle) by the fifth lumbar vertebra and by the last intervertebral disk.

The Osseous Structure of the Pelvic Girdle

The pelvic girdle has a vaultlike structure for which the flat bones participating in its formation provide increased structural strength. The pelvic girdle is composed of the two hip bones, which are the result of the fusion of the ilial bones, the ischial bones, the pubic bones, and of the sacrum, at which point the spinal column forms a firm connection with the pelvic girdle (Fig 3–8).

The pelvic girdle shows marked sexual

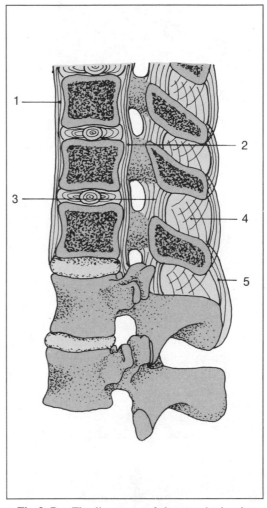

Fig 3–7.—The ligaments of the vertebral column (upper three vertebrae in cross section): *(1)* anterior longitudinal ligament, *(2)* posterior longitudinal ligament, *(3)* ligamentum flavum, *(4)* interspinal ligaments, *(5)* supraspinal ligaments.

differences, which are the result of the requirements of the childbearing process. The female pelvis is broader, roomier, less vertical, and is provided with a wider and deeper inlet and outlet.

The function of the pelvic girdle is to bear the weight of the thorax and to provide a connection for the lower extremity (Fig 3–9).

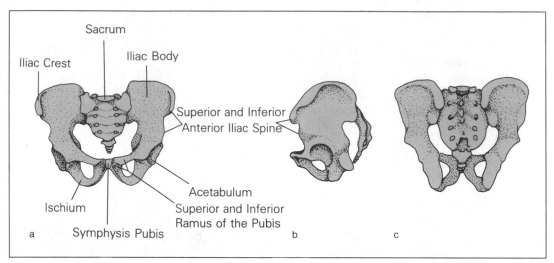

Fig 3–8.—Representation of the pelvic girdle: **(a),** anterior view, **(b),** lateral view, **(c),** posterior view.

The Articulations and Ligaments of the Pelvic Girdle

The hip bones are connected to the sacrum through the two sacroiliac joints and to each other through the cartilaginous *symphysis pubis*.

The ligaments of the pelvis are oriented according to the dominant stress lines and are stronger dorsally than ventrally (Fig 3–10).

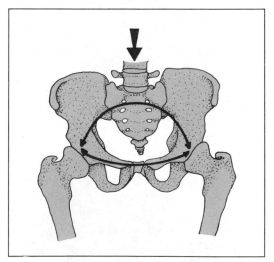

Fig 3–9.—The vault-shaped structure of the pelvic girdle and a diagrammatic representation of the distribution of pressure.

The Thorax

The thorax is located in the area of the spinal column between the head and neck area on one end and the lumbosacral area on the other.

The head-neck area is distinguished not only for its mobility, but also for its importance in performing gestures and expressions. The lumbosacral area has increasingly diminished mobility, but makes its contribution by making the erect posture and walking possible. The thorax lies between the two ends of the torso, which it bridges, from the perspective of locomotor dynamics. In addition, the thorax protects vital organs and supports the respiratory effort by rhythmic expansion and contraction.

The thorax is composed of the 12 thoracic vertebrae, 12 pairs of ribs, and the sternum. The first seven ribs have a direct cartilaginous connection with the sternum.

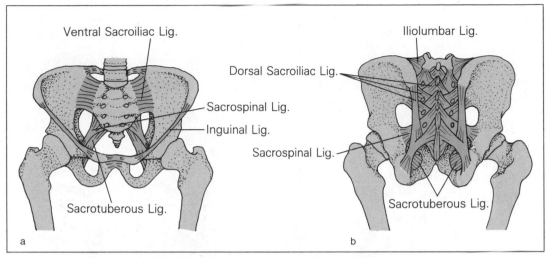

Fig 3–10.—The ligaments of the pelvic girdle: **a,** ventral, **b,** dorsal.

The lower five ribs are connected to the sternum through cartilaginous connections to the seventh costal cartilage, or not at all.

The mobility of the thorax is made possible by the costovertebral articulations, which allow a rotational movement of the ribs. This is important in allowing the expansion and constriction of the thorax, essential for respiration (Fig 3–11). The costal articulations will not be discussed in this volume.

The Respiratory Muscles and the Accessory Muscles of Respiration

Since breathing and, particularly, the various techniques of breathing are of paramount importance in all endurance tests and other cardiopulmonary stress situations, we will briefly review the different types of breathing—thoracic or abdominal—and the regular and accessory muscles of respiration.

Thoracic Breathing

In thoracic breathing, the ribs are raised during inspiration and lowered during expiration. As shown in Figure 3–11, both the sagittal and transverse diameter of the thoracic cavity is increased by raising the ribs. This is the inspiratory position. When the ribs are lowered, both diameters decrease. This is the expiratory position.

The True Respiratory Muscles

THE EXTERNAL INTERCOSTAL MUSCLES.— These muscles lie between the ribs in a posterosuperior to anteroinferior direction. Anteriorly, they end at the costochondral junction, beyond which they become membranous in nature. They originate on the external surface of the ribs, function as elevators of the ribs, and are, therefore, inspiratory muscles.

THE INTERNAL INTERCOSTAL MUSCLES.— These originate on the interior surface of the ribs and reach from the costal angle to the sternum. They cross the fibers of the *external intercostal muscles* almost at right

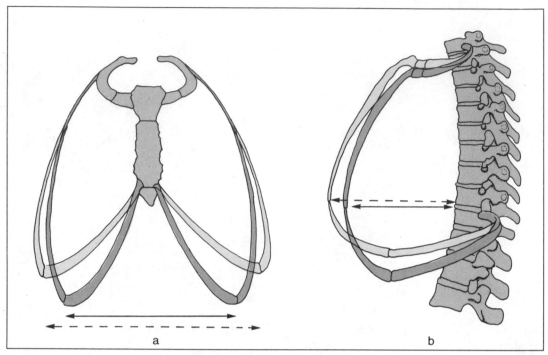

Fig 3–11.—**(a)** Anterior and **(b)** lateral view of the thorax in inspiration *(pink)* and expiration *(red)*. Increase in the transverse and sagittal diameter during inspiration.

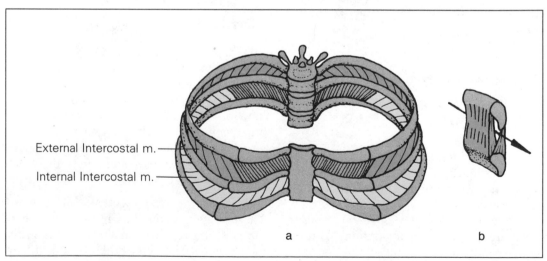

External Intercostal m.

Internal Intercostal m.

Fig 3–12.—**(a)** The external and internal intercostal muscles, **(b)** representation of the neurovascular space between the muscles.

angles. These muscles also have a membranous extension, at least at their posterior end.

Because of their orientation, the intercostal muscles can be considered as the continuation of the inner and outer abdominal musculature.

Between the two muscle layers we find a sheath containing nerves and blood vessels (Fig 3–12).

- *The transverse thoracic muscle*
- *The subcostal muscles*
  Both of these muscles are on the interior surface of the thoracic cage and serve an expiratory function.
- The *levatores costarum muscles*. These are muscles of inspiration.
- The *serratus posterior, superior, and inferior*. These muscles are to some extent respiratory, since the posterior *superior serratus m.* elevates the ribs and the *inferior serratus m.* lowers them.

In addition, they participate in the action of the *erector spinae muscle* and in the stretching motion of the trunk (Fig 3–13).

Note: In case of markedly increased demand (e.g., after a 400-m dash), thoracic breathing requires the assistance of the so-called *accessory muscles of respiration*. When the shoulder girdle is fixed (the arms are braced), all muscles that can elevate the ribs assist in inspiration. These are the *pectorales major and minor*, the *sternocleidomastoid*, and the *scalene muscles*. All muscles that lower the ribs can assist in expiration. These are the *iliocostals*, the *quadratus lumborum*, and the abdominal muscles.

Abdominal Respiration

In addition to *thoracic respiration,* there is also another type of breathing, the so-called *abdominal respiration*. This designation is based on the fact that the respiratory movements can be observed on the anterior abdominal wall.

The "motor" of *abdominal respiration* is

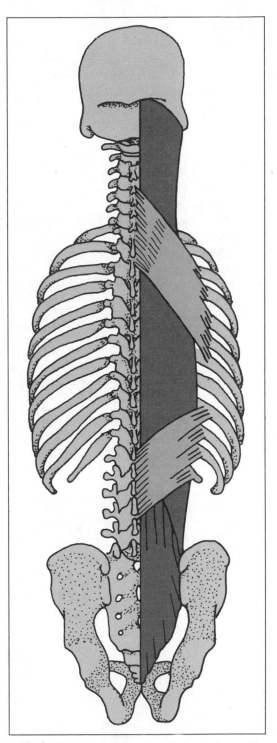

Fig 3–13.—The serratus posterior superior and inferior muscle (after *Rohen*).

the diaphragm. It originates from the sternum, from the internal surface of the six
lower ribs, and from the transverse processes of the first lumbar vertebra. It is
dome shaped and has its insertion in a central tendon (Fig 3–14).

A contraction of the diaphragm forces the
abdominal organs inferiorly, which leads to
a bulging of the anterior abdominal wall and
makes inspiration possible. When the abdominal muscles contract, they act as antagonists of the diaphragm, displace the diaphragm superiorly, and produce expiration.

Thoracic and abdominal respiration can
usually not be entirely separated and cooperate in varying degree.

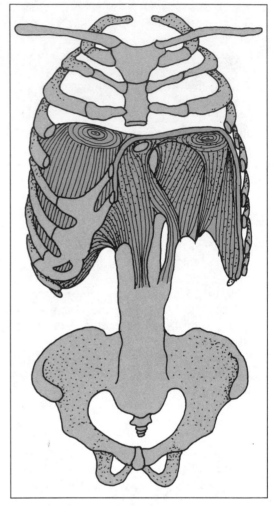

Fig 3–14.—The diaphragm and its component
parts.

The Active Locomotor System of the Trunk

The Abdominal and Dorsal Musculature

As already mentioned, the abdominal and dorsal musculature provides the trunk with a dynamic tension by virtue of the differential orientation of their muscle fiber bundles. This permits a wide range of complex motions.

The *abdominal musculature* has a number of *flat muscle sheets* which extend from the thorax to the upper limits of the pelvis and, which, among other functions, also protect the abdominal viscera. The *dorsal musculature,* on the other hand, is more *highly differentiated* and consists of a number of shorter and longer individual muscles.

The Abdominal Musculature

The Musculature of Anterior and Lateral Abdominal Wall

The lateral abdominal wall consists of three muscles. They are connected to the ventrally located, straight abdominal muscle by an aponeurosis. These aponeuroses surround the anterior muscles and contribute to their stability. In the back, they are anchored to the spinal column by the thoracolumbar fascia which surrounds the dorsal musculature anteriorly and posteriorly (Fig 3–15).

Rectus Abdominis m. (Fig 3–16)

Origin: 5–7 costal cartilages and the xyphoid process.
Insertion: Os pubis

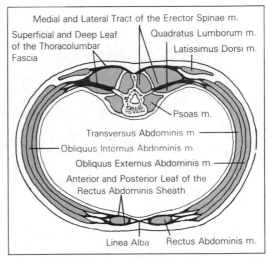

Fig 3–15.—The layers and attachment of the abdominal muscles represented at a cross section of the trunk. The "linea alba" is the site where the abdominal muscle sheaths join and cross over.

Innervation: Intercostal nerves
Function: If the pelvis is fixed, this muscle pulls the trunk forward (e.g., when performing "sit-ups" with the legs fixed). When the thorax is fixed, this muscle elevates the pelvis and in this function it is assisted by other muscles (e.g., swinging on the horizontal bar). If stimulated only on one side, the muscle bends the trunk laterally. It also serves to increase intra-abdominal pressure and assists in respiration (see p. 59).

The rectus abdominis muscle, by virtue of its insertion on the pubic bone, plays an important role in maintaining the level posture of the pelvis and, indirectly, the curvature of the lumbar spine.

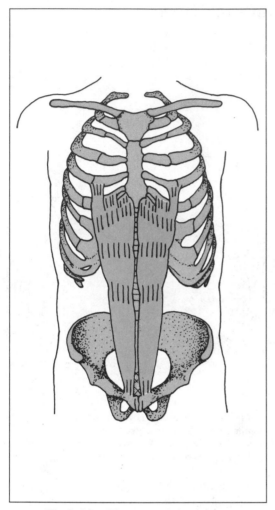

Fig 3–16.—The rectus abdominis m.

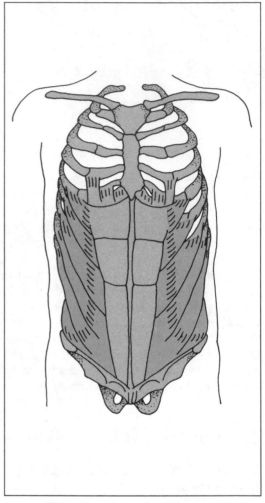

Fig 3–17.—The external oblique m.

If this muscle is poorly developed, the pelvis is tilted forward, which leads to increased lordosis of the lumbar spine and may lead to a posturally weak ''sway back'' condition.

External Abdominal Oblique m. (Fig 3–17)

Origin: External surface of ribs 5–12

Insertion: Iliac crest, inguinal ligament, pubic tubercle, and linea alba

Innervation: Intercostal, iliohypogastric, and ilioinguinal nerves

Function: On bilateral stimulation, these muscles assist the rectus abdominis in bending the trunk forward. If stimulated on one side only, they bend the trunk toward that side and rotate it in the opposite direction. In all track events involving throwing or tossing, these muscles participate in the terminal twisting motion of the trunk.

Internal Abdominal Oblique m. (Fig 3–18)

Origin: Iliac bone, inguinal ligament, lumbar aponeurosis

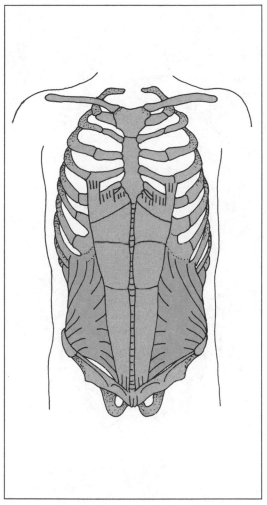

Fig 3–18.—The internal oblique m.

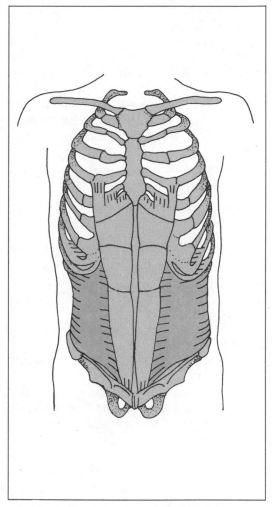

Fig 3–19.—The transversus abdominis m.

Insertion: Ribs 9–12, linea alba

Innervation: Same as the external oblique m.

Function: On bilateral stimulation, they bend the trunk forward. On unilateral stimulation, they bend the trunk sideways and rotate it toward the stimulated side. In lateral bending, the *internal oblique m.* cooperates with the ipsilateral *external oblique m.*, but in rotation with the contralateral *external oblique m.* The fibers of the *internal and external oblique muscles* cross at an angle of 90°.

Both muscles participate in increasing intra-abdominal pressure and in expiration.

Transversus Abdominis m. (Fig 3–19)

This muscle forms the deepest muscle layer of the abdominal wall.

Origin: Internal surface of costal cartilages 7–12, lumbar aponeurosis, iliac crest

Insertion: Linea alba

Innervation: Intercostal nerves, lumbar plexus

Function: The main function of this muscle is to increase intra-abdominal pressure.

It contributes, together with other muscles, to the molding of the waist.

An overview of the anterior and lateral abdominal muscles makes it obvious that these muscles enable the body to perform delicate and precise movements of the trunk. This is made possible by the fact that these large, poorly differentiated muscles run vertically, diagonally, and horizontally, and that they can pull in opposite directions through fibers that cross in the abdominal wall aponeurosis (see p. 61)

The Musculature of the Posterior Abdominal Wall

The posterior abdominal wall is composed primarily of the:

Quadratus Lumborum m. (Fig 3–20)

Origin: Iliac crest
Insertion: 12th rib, transverse processes of the lumbar vertebrae
Innervation: Intercostal nerves, lumbar plexus
Function: On bilateral stimulation, they pull the trunk backward and thus assist the *erector spinae m.* On unilateral stimulation, they pull the trunk to one side and, in conjunction with other muscles, assist in the precise positioning of the trunk.

Dorsal Musculature

The markedly sectionalized back muscles originate from the vertebral arches and vertebral processes. We can distinguish the flat superficial muscle layers, which act primarily on the upper extremity and shoulder girdle and which will be discussed later, and the true *autochtonous* back muscles, the primary function of which is the extension of the spinal column.

The Autochtonous Back Muscles

The *autochtonous back muscles* can be divided into *median* and *lateral* tracts. They

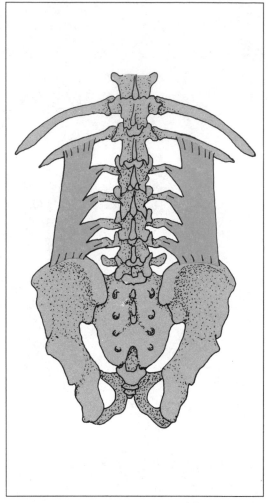

Fig 3–20.—The quadratus lumborum m.

are attached to the spinous and transverse processes by the thoracolumbar fascia. This arrangement prevents the muscles from pulling away from the back, like the string of a bow, whenever the back is extended.

The Median Tract

The median tract lies in the gutter between the spinous process and the transverse processes. It can be divided into three parts: the *spinal muscle* groups, which both originate and insert on the spinous processes; the *transversospinal* muscle group, which originates on the transverse processes and inserts on the spinous processes; and the *spinotransversal* muscle group, which originates on the spinous processes and inserts on the transverse processes.

In the *spinal muscle system,* we distinguish between the *interspinal muscles,* which bridge only neighboring spinous processes in the cervical and lumbar areas, and the spinalis muscles, which extend over several spinous processes, in the thoracic region. The shorter muscles lie deeper than the longer ones.

The *transversospinal system* (Fig 3–21) is divided, in ascending order, into:

* The *rotators breves et longi m.*
* The *multifidus m.*
* The *semispinalis m.,* which has a cervical and a thoracic segment

The *spinotransversal* system consists of the *splenius capitis* and *splenius cervicis* muscles (Fig 3–22).

Splenius capitis m.

Origin: Spinous processes of the upper thoracic (T_1–T_3) and lower cervical (C_3–C_7) vertebrae.

Insertion: Mastoid process and the superior nuchal line

Innervation: The dorsal branches of spinal nerves C_1–C_5

Function: When stimulated on one side, it turns the head toward the contracting muscle. If stimulated bilaterally, it draws the head backward.

Splenius Cervicis m.

Origin: The spinous processes of T_3–T_6

Insertion: The transverse processes of C_1–C_6

Innervation: As above

Function: When stimulated on one side, it bends the neck toward the contracting side. If stimulated bilaterally, it bends the

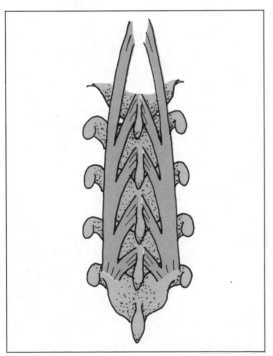

Fig 3–21.—Representation of the transversospinal system.

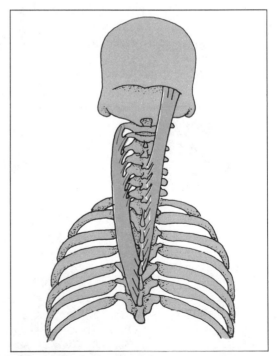

Fig 3–22.—The splenius capitis m. (**right**) and the splenius cervicis m. (**left**).

head backward. Through its insertion on the atlas, it also helps to turn the head.

The Lateral Tract

While the *median tract* consists mostly of short muscles, the *lateral tract* is composed mostly of long muscles. The lateral tract, which unites in the lumbar region into a powerful muscular tract, the *erector spinae muscle,* is composed of two parts: the *longissimus* and *iliocostal muscles* (Fig 3–23).

Longissimus m.

Origin: This muscle consists of a capitis, cervicis, and thoracis part. It accordingly originates from the transverse processes of all the vertebrae, the sacrum, and the iliac crest.

Insertion: The *capitis* portion inserts on the mastoid process, the *cervicis* portion inserts on the transverse processes of the cervical vertebrae, and the *thoracis* portion inserts on the ribs and on transverse processes of the thoracic and lumbar vertebrae.

Innervation: The dorsal rami of appropriate spinal nerves

Function: Stimulation of one side results in the lateral flexion of the particular spinal column segment. Bilateral stimulation results in an extension of the back.

Iliocostalis m.

Origin: This muscle also has three components: the cervical, thoracic, and lumbar. They originate respectively from the third to 12th rib, the sacrum, and the iliac crest.

Insertion: All 12 ribs and the transverse processes of the cervical vertebrae.

Innervation: As above

Function: Similar to the *longissimus m.,* it serves to extend and laterally bend the trunk. It also pulls down on the ribs and thus assists in expiration (see p. 59).

If the *erector spinae m., (erector trunci)* is weak, it may lead to a postural defect, the so-called ''humpback.''

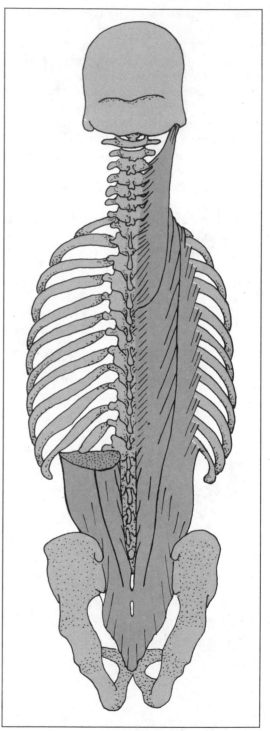

Fig 3–23.—The two components of the erector spinae m.

Hanni Wenzel on her way to
victory in World Cup slalom.

The abdominal and dorsal musculature provides the trunk with dynamic tension that can adapt itself, in a most sophisticated way, to the manifold motions of the trunk and of the extremities. The primary function of these muscles is the maintenance of the erect posture.

The spinal column in this respect resembles the mast of a sailboat that has a system of braces and that is anchored in a vertical position on the keelson (pelvis).

Variations in tension in one direction must be accompanied by variations in the entire system of braces. In the same way, the abdominal and dorsal muscles can never function in isolation, but always only as a complete system.

General Overview of the Function of the Abdominal and Dorsal Musculature

The simple diagrammatic representation emphasizes the differences in the structure of the anterior and posterior tension system, which are shown in a somewhat distorted proportion for the sake of clarity. In fact, the abdominal musculature is considerably less structured than the dorsal musculature (Fig 3–24).

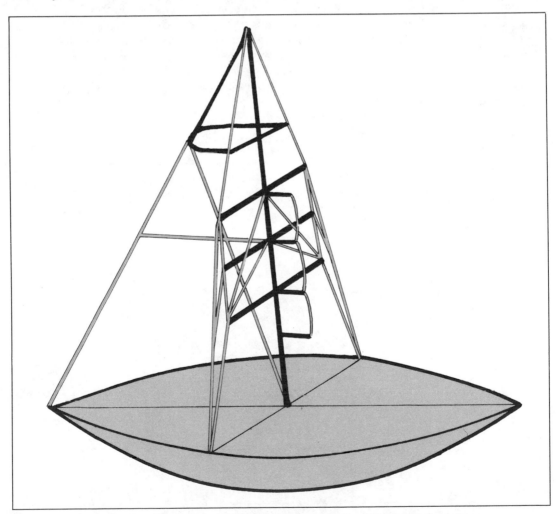

Fig 3–24.—Ship's mast model of the bracing scheme of the vertebral column.

The Articulations Between the Spinal Column and the Head

In contrast to the relatively immobile lumbar spine, which is solidly anchored to the pelvic girdle, the cervical spine has an extraordinary mobility in all directions. This is most important, since the head, which contains the sense organs, is attached to the cervical spine. The mobility of the head is made possible through some very special articulations (Fig 3–25).

There is an atlanto-occipital joint, between the condyles of the occipital bone and the corresponding articulating surfaces of the atlas. There is also a group of atlantoaxial joints between the first two cervical vertebrae.

The atlanto-occipital articulation allows for extension and flexion of the head and for limited lateral motion. The atlantoaxial joints allow the horizontal turning motions of the head.

As shown in Figure 3–26, the first two cervical vertebrae differ in design from all the other vertebrae. As part of human evolution, the atlas has given up its vertebral body to the axis and the "tooth" of the axis (dens) is the result of this transfer. The atlas has, thus, been transformed into a ring that can rotate around the dens of the axis.

The Ligamentous System of the Head

To keep the dens of the axis from penetrating the proximal end of the spinal cord during head and neck motion it is fixed in place by a series of ligaments. The *ligamentum transversum* is the only one of numerous ligaments that will be mentioned here. It secures the dens of the axis in the anterior arch of the atlas, and participates in the formation of the middle atlantoaxial joint, by virtue of a cartilaginous layer on its inner surface.

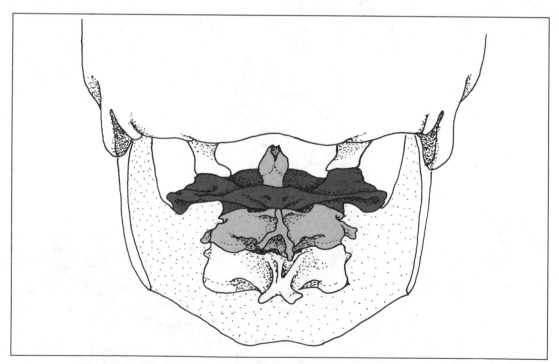

Fig 3–25.—Representation of the craniovertebral attachment and its articulations.

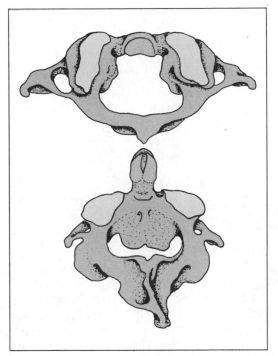

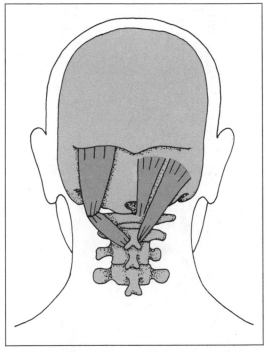

Fig 3–26.—Representation of the atlas **(above)** and axis **(below)**.

Fig 3–27.—The muscle system of the articulations of the head. **Left,** obliquus capitis superior and inferior m.; **right,** rectus capitis posterior minor and major m.

The Muscles Affecting the Position of the Head

As already mentioned, the head is attached to the cervical spine in such a fashion as to allow maximum mobility. Such mobility is important, both for spacial orientation and also for the creation of individual expressions.

The delicate positioning of the head is accomplished by a multicomponent system of small muscles that will be shown graphically, but not discussed in detail (Fig 3–27).

In the area of athletics, the delicate positioning of the head is important not only for spacial orientation, as, for example, in diving and trampoline jumping, but also for a fixation of the head as, for example, in soccer. This fixation of the head is accomplished by a group of larger and stronger muscles that act by the coordination of the

isometric tension of all the muscles acting on the articulations of the head. Only the most important ones will be discussed (Fig 3–28).

In the region of the nape of the neck, only the trapezius muscle (see p. 74) and the deep components of the median and lateral bundles of the autochtonous muscles of the back (*semispinalis capitis, splenius capitis,* and the capitis portion of the *longissimus m.*) need to be mentioned.

In the lateral and anterior region, the *sternocleidomastoid muscle* is dominant.

Sternocleidomastoid m. (Fig 3–29)

Origin: Sternum and clavicle

Insertion: Mastoid process and linea nuchae superior

Innervation: The accessory nerve and branches of the cervical plexus

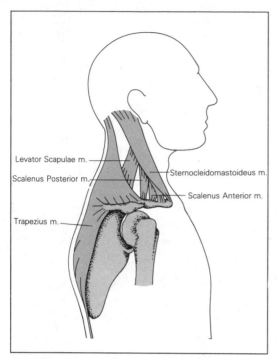

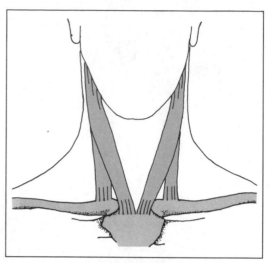

Fig 3–29.—The sternocleidomastoid m.

Function: In bilateral stimulation, it forcefully bends the head forward and acts in opposition to the trapezius m. When stimulated on one side only, it bends the head forward and turns it toward the opposite side. It assists in inspiration.

Fig 3–28.—Overview of the most important neck muscles.

Levator Scapulae m.

Scalenus Posterior m.

Trapezius m.

Sternocleidomastoideus m.

Scalenus Anterior m.

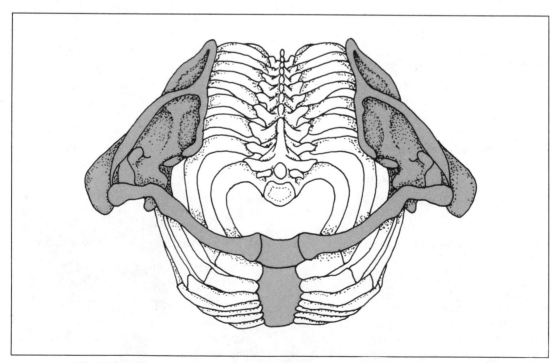

Fig 3–30.—Diagrammatic representation of the shoulder girdle (view from above).

The *scalenus muscles,* acting on the cervical spine, also affect the position of the head. They are partially covered by the *sternocleidomastoid m.* There is an anterior, middle, and posterior scalene muscle. They originate on the transverse processes of the cervical vertebrae and insert on the first and second rib. When stimulated on one side, they bend the cervical spine laterally. When stimulated bilaterally, they elevate the thoracic cage and assist in inspiration.

The Upper Extremity

The Shoulder Girdle

The *shoulder girdle* attaches the arm to and connects it with the trunk. It is composed of the scapula, clavicle, and sternum (Fig 3–30).

Contrary to the *pelvic girdle,* which is relatively firmly attached to the spinal column and lower extremity, the attachments of the *shoulder girdle* are very loose.

The mobility of the shoulder girdle almost doubles the range of motion of the upper extremity. This is of great importance in extending the grasping range of the hands (Fig 3–31).

The Articulations of the Shoulder Girdle

The shoulder girdle forms a functional unit, which is made possible by two ball and socket joints and the internal and external clavicular joints (Fig 3–32).

The *internal clavicular joint* forms a union between the clavicle and the sternum. This joint is the only osseous connection between the shoulder girdle and the thorax. It braces the shoulder girdle against the thorax. The mobility of this joint is severely limited by tight ligaments (see Fig 3–32).

The *external clavicular joint* is composed by the clavicle and acromion. In this joint, motion is also severely limited by the surrounding tendons.

The Musculature of the Shoulder Girdle

The function of these muscles is to fix the shoulder girdle to the trunk and to place the shoulder girdle in a suitable position for the various different movements of the upper extremity.

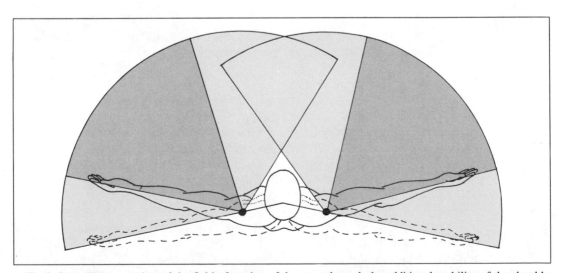

Fig 3–31.—The expansion of the field of motion of the arms through the additional mobility of the shoulder girdle (expanded field in red).

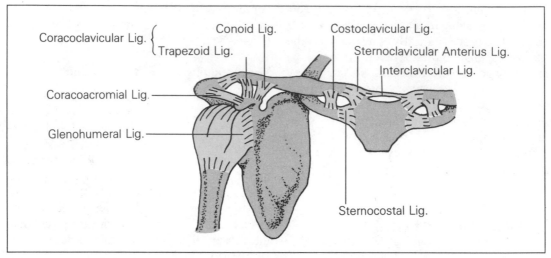

Fig 3–32.—Osseous and ligamentous system of the shoulder girdle (after *Benninghoff*).

Trapezius m. (Fig 3–33)

The *trapezius m.*, together with the *latissimus dorsi m.*, cover almost the entire back (see also p. 77).

Origin: The occipital tuberosity, the spinous processes of the cervical and thoracic vertebrae.

Insertion: Clavicle, the acromion, and the spine of the scapula.

Innervation: The accessory nerve and branches of the cervical plexus.

Function: According to the orientation of its component fibers, the *trapezius m.* can be divided into a superior, middle, and inferior portion.

The *superior part* raises the shoulder and assists in rotating the scapula (see p. 137). It plays an important role in all pulling and lifting motions, and is, hence, particularly well developed in weight lifters. When stimulated on one side only, the superior portion of the *trapezius m.* rotates the head in the opposite direction. The clavicular portion raises the clavicle and assists in inspiration.

The *middle, horizontal part* pulls the scapula toward the spinal column (e.g., when the arms are pulled posteriorly).

The *inferior portion* pulls the shoulder down, and jointly, with the superior portion, assists in rotating the scapula. In support beam exercises, the trapezius m., with the assistance of other muscles, prevents the sagging of the trunk.

The *trapezius m.* is rarely used in its entirety, and its component parts usually cooperate, individually, with other muscle groups. This muscle is a good illustration of the fact, that the compound muscles can perform a variety of functions, even occasionally opposing movements, depending on the origin and insertion of their different components.

Rhomboideus Major M. (Fig 3–34)

This muscle and those to follow, all lie below the *trapezius m.*

Origin: The spinous process of the first four thoracic vertebrae

Insertion: Medial edge of the scapula

Innervation: The dorsalis scapulae n.

Rhomboideus Minor m. (see Fig 3–34)

Origin: The spinous process of two lower cervical vertebrae.

Insertion: As above

Innervation: As above

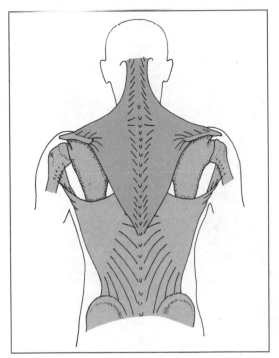

Fig 3–33.—The trapezius m.

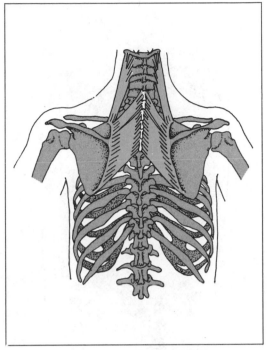

Fig 3–34.—Representation of the rhomboideus major m. **(below),** minor m. **(middle),** and the levator scapulae m. **(above).**

Function: *Both* muscles elevate the scapula toward the spinal column. The *rhomboideus major* m. is also important in rotating the inferior angle of the scapula in direct opposition to the *serratus anterior m.* (see below).

Levator Scapulae m. (see Fig 3–34)

Origin: The transverse processes of the first four cervical vertebrae
Insertion: Superior angle of the scapula
Innervation: The dorsalis scapulae n.
Function: It pulls the scapula upward (e.g., when shrugging the shoulder). It always functions in conjunction with other muscles.

Serratus Anterior m. (Fig 3–35)

Origin: Ribs 1–9
Insertion: Medial edge of the scapula and both the superior and inferior angle of the scapula
Innervation: The long thoracic n.

Function: In its entirety, this muscle immobilizes the scapula against the trunk. Its *superior portion* elevates the scapula; its *middle portion* functions in opposition to the horizontal portion of the *trapezius m.* The *inferior portion* is particularly important in rotating the scapula. It pulls the inferior angle of the scapula forward and so makes it possible to raise the arms above the horizontal (see p. 137).

When the arms are braced and the scapula immobilized, the *serratus anterior m.* elevates the ribs and assists inspiration.

Pectoralis Minor m. (Fig 3–36)

Origin: Ribs 2–5
Insertion: Coracoid process
Innervation: The median and lateral pectoralis n.
Function: This muscle lying under the

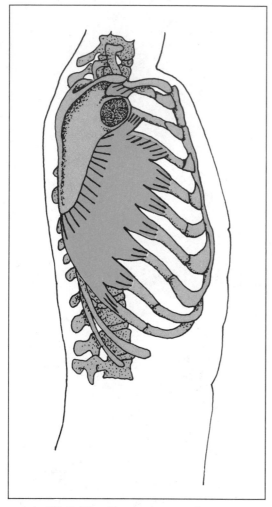

Fig 3–35.—The serratus anterior m.

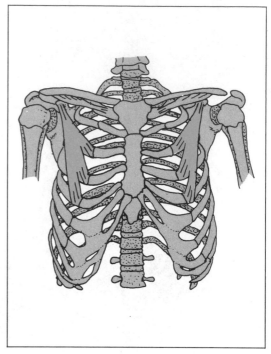

Fig 3–36.—The pectoralis minor m.

The Shoulder Joint

The two bones participating in the *shoulder joint* are the humerus (head) and the scapula (glenoid cavity).

Because of the difference in the articulating surface area between the scapula and the humerus (1:4 ratio) and because of the looseness of the joint capsule, the *shoulder* is the most freely movable joint in the body, and also the one most prone to dislocation.

Because of the construction of the shoulder joint, anterior and lateral motion of the arm is impeded at or about the level of the shoulder and when the arm reaches a horizontal position. This limitation of motion can be compensated for by rotating the scapula, which allows the raising of the arm above the shoulder and into the vertical position (see Fig 3–32).

Clinical reference: When the arm is hanging loosely, the joint capsule becomes

pectoralis major (see p. 59) lowers the shoulder girdle.

When the shoulder girdle is immobilized, it can elevate the ribs and assist in inspiration (see p. 78). This muscle rarely acts in isolation.

Sternocleidomastoid M. (see Figs 3–28 and 3–29).

This muscle originating on the skull, also acts on the shoulder girdle (see p. 59 for discussion).

pleated. Because of this, the arm must be splinted in an abducted and anterior position, whenever long immobilization becomes necessary (e.g., fractures). If this is not done, the folds in the capsule may fuse, or the capsule may shrink, events that both will inevitably lead to a significant limitation of motion.

The Muscles of the Shoulder Joint

Latissimus Dorsi m. (Fig 3–37)

This muscle is superficial and, together with the *trapezius m.*, covers almost the entire back. It is one of the largest flat muscles of humans. It forms the posterior axillary fold together with the *teres major m.*

Origin: Through the thoracolumbar fascia from the spinous process of all thoracic and lumbar vertebrae, the sacrum, and the iliac crest

Insertion: On the ridge of the lesser tubercle of the humerus

Innervation: The thoracodorsalis n.

Function: The *latissimus dorsi m.* rotates the arm inward, and pulls it backward. It pulls the raised arm downward with great force and this makes it a very important

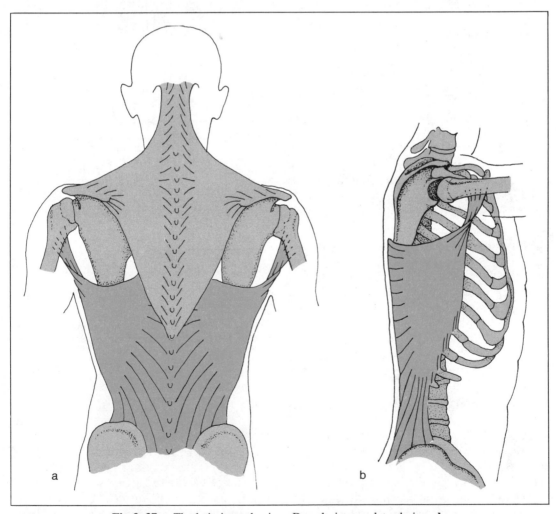

Fig 3–37.—The latissimus dorsi m. Dorsal view, **a,** lateral view, **b.**

muscle in all throwing and hitting motions. When the arm is extended laterally, it serves as an adductor (e.g., in the "iron cross hang" on the still rings). When the arms are fixed, it raises the body (e.g., chin-up exercises). In the long hang swing on the horizontal bar or in the arm support position on the parallel bars, it prevents the downward displacement of the trunk. In this function, it is assisted by the *pectoralis major m.* These two muscles hold the trunk in a muscular support sling (Fig 3–38).

Pectoralis Major m. (Fig 3–39)

This muscle forms the anterior axillary fold.

Origin: Clavicle, sternum, and the fascia of the abdominal wall

Insertion: The ridge on the greater tubercle of the humerus

Innervation: The anterior pectoral n.

Function: Since the *pectoralis major m.* has descending, horizontal, and ascending fibers, there are different functions attributable to these components. The only time that all fibers are aligned and function in the same general direction is when the arm is forcibly lowered from an overhead position (e.g., javelin throw). When the arm is raised, but fixed (e.g., pole vaulting) the pectoralis major, together with other muscles, raises and turns the trunk. Its function as a sling, to immobilize the trunk, has been described above (see Fig 3–38).

In most contractions of the *pectoralis major m.*, one or the other component tends to dominate to a greater or lesser degree. The horizontal sternocostal portion dominates when the dorsolaterally extended arm is forcefully brought forward (e.g., discuss throw). When the arm is extended downward and backward, the clavicular portion is used to bring the arm forward (e.g., bowling). Both the clavicular and horizontal parts participate in the inward rotation of the arm. Adduction is accomplished primarily by the ascending and horizontal portions. When the arms are braced, the *pec-*

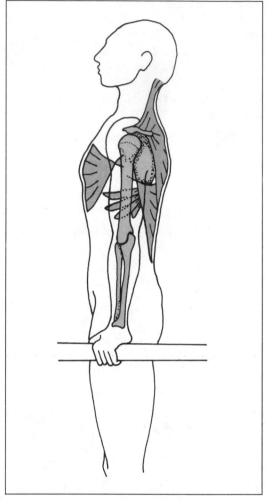

Fig 3–38.—The support sling of the trunk that prevents the downward displacement of the trunk in all hanging and support exercises. The latissimus dorsi m., the pectoralis major, and minor m., and the inferior portion of the serratus anterior m. participate in connecting the trunk to the shoulder girdle.

toralis major m. assists in inspiration (see p. 59).

Note: In order that the three components of the muscle (all of which have fibers going in different directions and being of different lengths) do not hamper the forward and lateral elevation of the arm, the pectoralis major m. compensates by having unusually elastic fibers and by an arrangement of its insertion tendons, which by criss-

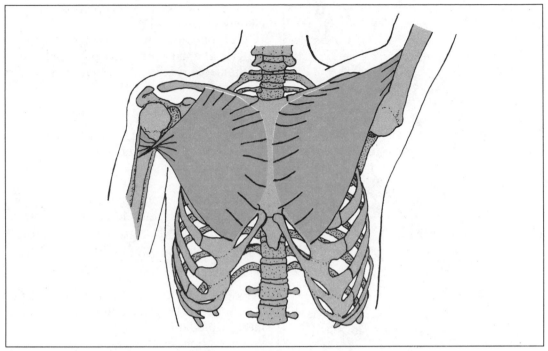

Fig 3–39.—The pectoralis major m.

crossing make up for the differing length of the fibers. The fibers, which originate on the upper part of the trunk, insert as the deepest component; those that originate at the lower levels, insert as the top-most compartment. This particular insertion arrangement gives the muscle a certain tone in all possible positions and, also, prevents the overextension of the ascending fibers when the arm is raised (Fig 3–40).

Deltoid m. (Fig 2–7 and 3–41)

This muscle has several components that not only participate in *all* movements of the shoulder joint and thus provide an extraordinarily important dynamic function, but that also have an important role in stabilizing the joint itself. It covers the joint like a hood and is a major factor in holding the joint together. This muscular protection explains why most dislocations of the joint take place in a downward direction, where no such protective muscle exists.

Whenever the *deltoid m.* atrophies (e.g.,

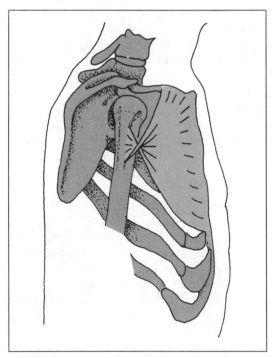

Fig 3–40.—The insertion of the pectoralis major m. tendon (after *Rohen*).

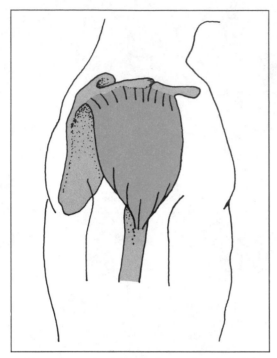

Fig 3–41.—The deltoid m. (see also Fig 2–7).

nerve injury), the shoulder joint becomes very loose.

Origin: The three components originate from the scapula, the acromion, and the spine of the scapula, respectively.

Insertion: The deltoid tuberosity of the humerus.

Innervation: The axillary n.

Function: The *anterior* portion moves the arm forward, the *posterior* portion moves the arm backward, and the *middle* portion moves the arm laterally. In addition, the *anterior* portion rotates the arm internally and the *posterior* portion externally. The *deltoid m.* is the most versatile muscle of the shoulder joint. It is the swimming muscle above any other. At the crawl, for instance, this is the muscle that raises the lead arm, rotates it and brings it forward. It is particularly strongly developed in weight lifters, because of its importance in abduction.

Since this muscle participates in all movements of the shoulder joint it is in it-self a complex system of synergists and antagonists (see p. 42).

The following muscles all originate on the scapula and go the the proximal end of the humerus. Figure 3–42 gives a summary overview of these muscles.

Supraspinatus m. (Fig 3–43)

Origin: Supraspinous fossa of the scapula
Insertion: The greater tubercle of the humerus and upper surface of external rotator cuff.
Innervation: Suprascapular n.
Function: This muscle assists the *deltoid m.* in abducting the arm and externally rotating it.

Infraspinatus m. (see Fig 3–43)

Origin: Infraspinous fossa of the scapula
Insertion: The greater tubercle of the humerus (middle facet)
Innervation: Suprascapularis n.
Function: The upper fibers abduct the arm; the lower fibers adduct it. The *infraspinatus m.* has the major role in the backward extension and external rotation of the arm (wind up before a pitch), and is capable of developing more force in this external rotation than any other muscle.

Teres Minor m. (Fig 3–44)

Origin: Scapula
Insertion: The greater tubercle of the humerus (inferior facet)
Innervation: Axillary n.
Function: It adducts the arm, rotates it externally, and lowers the raised arm backward. (Other muscles participate in this movement.)

Teres Major m. (see Fig 3–44)

Origin: The inferior, lateral part of the scapula
Insertion: The lesser tubercle of the humerus
Innervation: The subscapular nerves
Function: It adducts the arm and, in op-

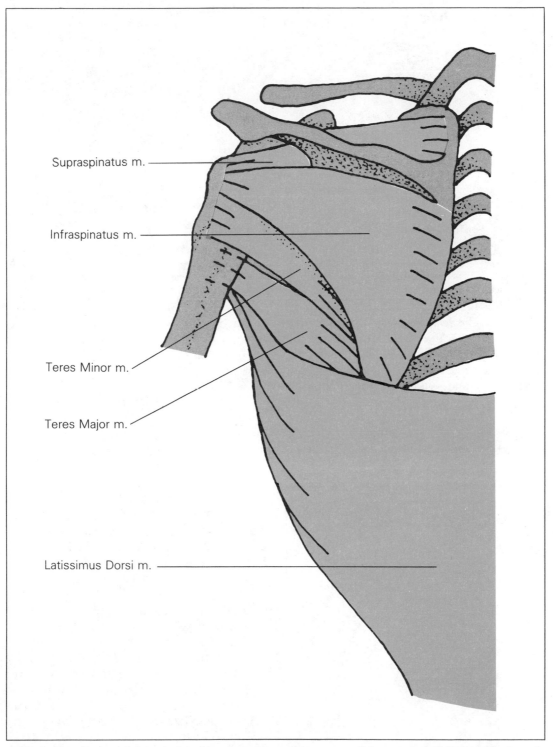

Fig 3–42.—Survey of the muscles going from the posterior aspect of the scapula and from the back to the arm.

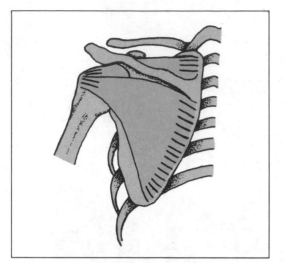

Fig 3–43.—The supraspinatus m. (**above**) and the infraspinatus m. (**below**).

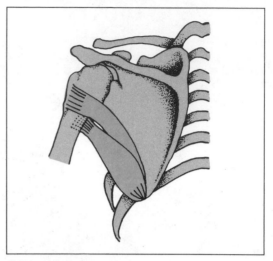

Fig 3–44.—The teres major m. (**below**) and the teres minor m. (**above**).

position to the *teres minor m.*, rotates it internally. It pulls the raised arm forward and downward (e.g., in free-style swimming and cross-country skiing). It also assists in all throwing and striking motions. When the arms are fixed, it pulls the trunk toward the arms (e.g., front support swing on the horizontal bar). It thus resembles the *latissimus dorsi m.* in a number of functions.

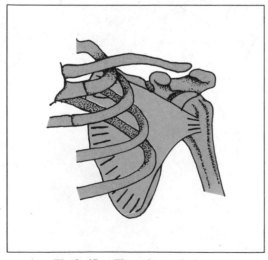

Fig 3–45.—The subscapularis m.

Finally, the anterior surface of scapula (this is the side toward the ribs) serves as the origin of the *subscapularis m.*

Subscapularis m. (Fig 3–45)

Origin: The costal surface of the scapula
Insertion: The lesser tubercle of the humerus
Innervation: The subscapular nerves
Function: This muscle rotates the arm inward and pulls the raised arm downward. It thus also participates in throwing and striking motions. Its *lower* fibers adduct the arm, its *upper* fibers abduct it. In walking, it assists in the pendulumlike backward-and-forward swinging of the arms.

Coracobrachialis m. (Fig 3–46)

Origin: The coracoid process
Insertion: Anterior and interior surface of the humerus
Innervation: Musculocutaneous n.
Function: This muscle adducts the raised arm and rotates it internally. In walking, it assists in swinging the arms forward. It also helps to stabilize the shoulder joint.
The following two muscles are biarticular, which act both on the shoulder joint,

The muscles of the arms and
the anterior and posterior
aspect of the trunk.

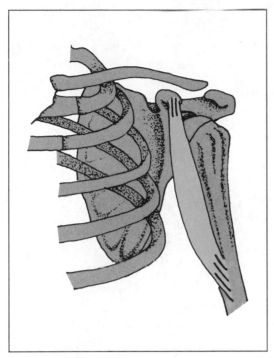

Fig 3–46.—The coracobrachialis m.

but, more particularly, on the elbow joint. They will be discussed in more detail below.

Biceps Brachii m. (see Fig 3–49)

Function at the shoulder joint: the *long head* raises the arm from the dependent position and rotates it internally. The *short head* adducts the arm and swings it forward in walking, very much like the *coracobrachialis m.*

The long tendon of the biceps brachii m. transverses the joint capsule and originates on the upper lip of the glenoid fossa. The *biceps brachii m.* thus also participates in stabilizing the shoulder joint.

Triceps Brachii m. (see Fig 3-53)

Function at the shoulder joint: the *long head* pulls the raised or anteriorly extended arm downward and backward as in swimming (all forms), all throwing or striking motions, and in cross-country skiing. When the arms are fixed, it pulls the trunk toward

the arms. In this function, it cooperates with a number of other muscles.

The Elbow Joint

The elbow joint connects the arm with the forearm. It is a *triple joint,* which is arranged in such a fashion that the three component joints can perform both rotational and hinge movements simultaneously, in all positions, and independently of each other.

This makes it possible for the most important part of the upper extremity, the hand, to function in the widest possible range of holding, grasping and pointing.

The Osseous and Ligamentous Structure of the Elbow Joint

The elbow joint is composed of three distinct articulating components. The humerus, the ulna, and the radius articulate with each other inside a single joint capsule (Fig 3–47).

These three joints can be identified in Figure 3–48.

The Joint Between the Humerus and the Ulna

The articulating surfaces are the trochlea of the humerus and the trochlear notch of the ulna, which holds the trochlea almost like the jaws of a pair of pliers. This joint allows only a single-axis, hinge motion.

The Joint Between the Humerus and the Radius

Although this joint is anatomically a ball and socket joint, the very strong collateral ligaments allow only rotating and hinge motions. During rotation, the head of the radius turns on its own axis within the annular ligament.

The Joint Between the Ulna and the Radius

In this joint, the head of the radius rotates in the corresponding radial notch of the ulna.

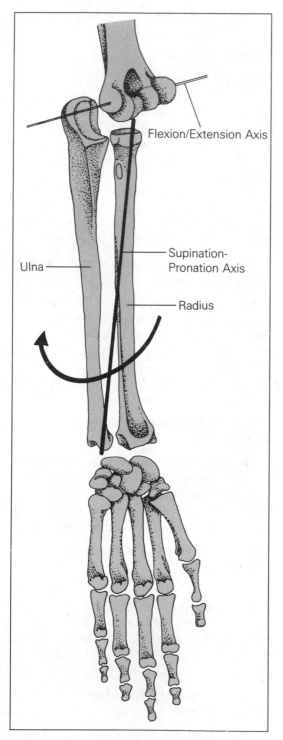

Flexion/Extension Axis

Supination-
Pronation Axis

Ulna

Radius

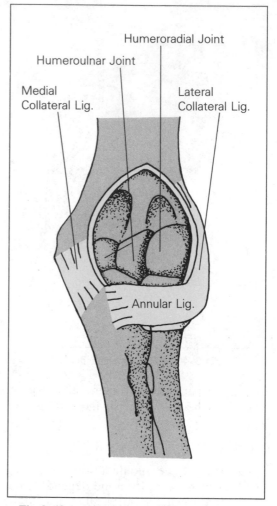

Humeroradial Joint

Humeroulnar Joint

Medial
Collateral Lig.

Lateral
Collateral Lig.

Annular Lig.

Fig 3–48.—Articulations and ligaments of the elbow joint (after *Benninghoff*).

Fig 3–47.—The axes of rotation of the elbow joint
represented on the left arm.

The Muscular System of the Elbow Joint

The Muscles of Flexion and Extension

Since the different flexed positions are much more important in the manifold grasp, touch, and expressive functions of the hand, it is easily understandable why the muscles of flexion are more structured than the muscles of extension.

Flexors

Biceps Brachii m. (Fig 3–49)

Origin: The short head from the coracoid process, the long head from the supraglenoid tubercle
Insertion: Radial tuberosity
Innervation: Musculocutaneous n.
Function: As a biarticular muscle, the *biceps brachii m.* affects both the shoulder joint (see p. 85) and the elbow joints.

At the elbow, the biceps brachii m. bends the forearm (e.g., chin-ups) and turns it from the pronated position into the supinated position (palms up) as in the breast stroke between the forward thrust and the lateral pull. The muscle can develop its fullest power in 90° flexion and in supination (chin-ups are easier with flexed elbows than with the arms straight) since in this position the tendon is not wound around the radius and the ordinarily rotated muscle is brought into its optimal pull position. For gymnasts this means, however, that they must train the *biceps brachii m.* both in the overgrip and undergrip positions to be prepared for both types of grips, each of which uses different fiber groups of the same muscle.

Because of its short lever action, the *biceps brachii m.* is a typical speed-lifter, and even a minimal shortening of the fibers causes a marked displacement of the hands.

Brachialis m. (Fig 3–50)

Origin: It lies under the biceps and originates from the anterior surface of the distal humerus.

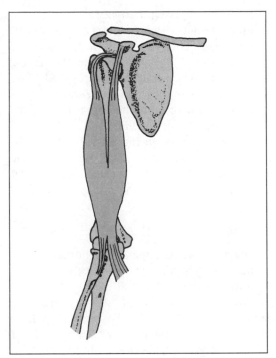

Fig 3–49.—The biceps brachii m.

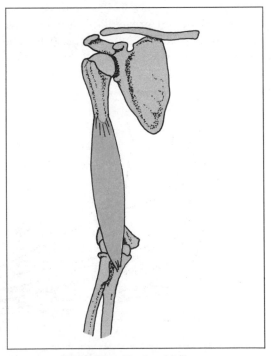

Fig 3–50.—The brachialis m.

Insertion: Ulnar tuberosity

Innervation: Musculocutaneous n.

Function: This muscle is a pure forearm flexor, almost as powerful as the *biceps brachii m.* When the forearm is fixed (e.g., long hang swing on the horizontal bar), it pulls the arm toward the forearm and thus assists in performing chin-ups.

Since it inserts on the ulna, while the *biceps brachii m.* inserts on the radius, the brachialis m. provides for a better distribution of load between the bones of the forearm.

Brachioradialis m. (Fig 3–51)

Origin: Lateral edge of the humerus

Insertion: Styloid process of the radius

Innervation: The radial n.

Function: Because of its long lever action, this muscle, which lies primarily in the forearm, is a typical heavy load flexor. It effects its greatest flexion strength, contrary to the *biceps brachii m.,* in the pronated position.

As shown in Figure 3–52, the three flexor muscles can exert greater or lesser leverage on the forearm. The *biceps brachii m.* and the *brachialis m.* have short lever action and can, therefore, be classified as *rapid flexors.* The *brachioradialis m.,* on the other hand, has a long lever action and is therefore known as a typical *heavy load flexor.* Since the ratio between the power lever and the weight arm is 1:5, for every load of 1 kg, a flexion force of 5 kg must be generated. In the *brachioradialis m.,* the two lever arms are of almost identical length and thus for the same load only one fifth of the force must be generated by this muscle. This illustrates the point that for different muscular functions there are corresponding muscular arrangements, and also, that a staggering of the lever action allows for a much broader utilization of the entire muscle system.

Extensors

The elbow joint has only one extensor:

Triceps Brachii m. (Fig 3–53)

Origin: The long head originates from the infraglenoid tubercle; the middle and lateral heads originate from the posterior aspect of the humerus.

Insertion: The olecranon

Innervation: The radial n.

Function: Extension of the elbow joint. This muscle plays an extraordinarily impor-

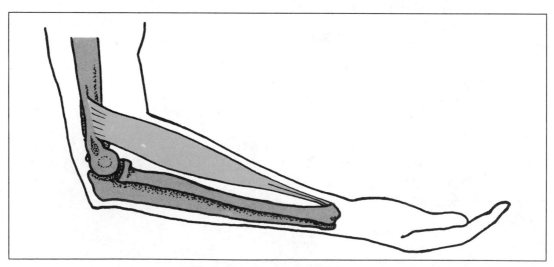

Fig 3–51.—The brachioradialis m.

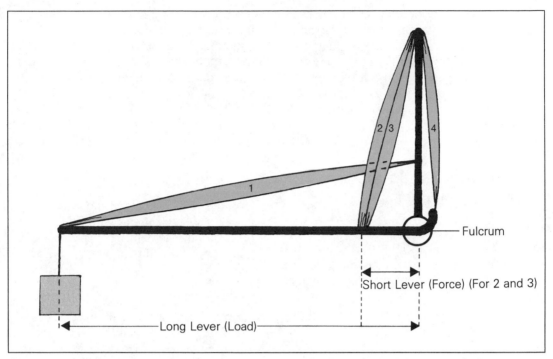

Fig 3–52.—The short (force) and long (load) levers of the flexors of the elbow. **(1),** The brachioradialis m. **(2),** the biceps brachii m. **(3),** the brachialis m. **(4),** the triceps brachii m. For the brachioradialis m. the two levers are of approximately the same length.

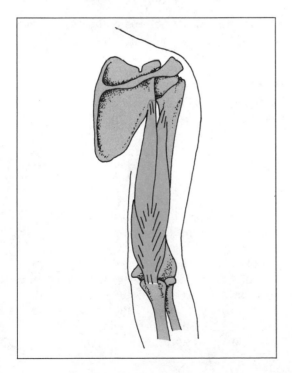

Fig 3–53.—The triceps brachii m.

tant role in a variety of athletic activities. In every case, where the extension or fixation of the elbow joint is desirable, the strength of the *triceps brachii m.* is the limiting factor (e.g., shot-put, boxing, all gymnastic apparatus, and weight lifting).

The function of this muscle at the shoulder joint was discussed above.

The Rotational Joints of the Forearm

Pronation (rotation of the back of the hand upward or inward) and *supination* (rotation of the palm upward or outward) are performed by two anatomically separate joints: the *proximal* and *distal radioulnar articulations.* Since these two joints form a functional unit, they will be discussed together.

As shown in Figures 3–54 and 55, pronation causes the radius to flip over the ulna in a diagonal direction. In other words, the pronation-supination axis extends from the head of the radius, at an angle across the forearm, to the head of the ulna (see Fig 3–47).

Pronation and supination can be performed more efficiently when the elbow is straight (230°), than when the elbow is flexed (130°), since the motion of shoulder joint contributes to pronation-supination when the arm is extended.

Muscles That Act on the Rotational Joints of the Forearm

Pronators

Pronator Teres m. (see Fig 3–54)

Origin: Medial epicondyle of the humerus and the coronoid process of the ulna.

Insertion: Middle third of the radius
Innervation: Median n.
Function: Because of its course, anterior to the axis of the elbow joint, to the radius, the *pronator teres m.* not only rotates the forearm interiorly, but is also a powerful flexor of the forearm. This twofold function is the reason why this muscle plays a role in the development of the so-called, javelin thrower's elbow (see p. 162).

Pronator Quadratus m.

Origin: Distal quarter of the ulna
Insertion: Anterior surface of the radius
Innervation: Median n.
Function: In conjunction with the pronator teres m., this muscle turns the hand inward.
Note: Pronation is more powerful when the arm is straight, since in this position the *pronator teres m.* is slightly stretched and thus can generate greater force.

Supinators

Supinator m. (see Fig 3–55)

Origin: The lateral epicondyle of the humerus, lateral collateral ligament, annular ligament, and ulna
Insertion: Middle portion of the radius
Innervation: The radialis n.
Function: As its name suggests, this muscle supinates the forearm. In addition, it extends the elbow joints, since a portion of its fibers lie posteriorly to the axis of the elbow joint.

Biceps Brachii m.

This muscle, discussed in detail above, is the primary muscle of supination. Its power to supinate is three times greater than the supinator muscle.

Note: Contrary to the pronators, the supinators are most powerful when the elbow joint is flexed. This is due to two reasons: (1) the biceps brachii m. can exert its maximal power when the elbow joint is flexed,

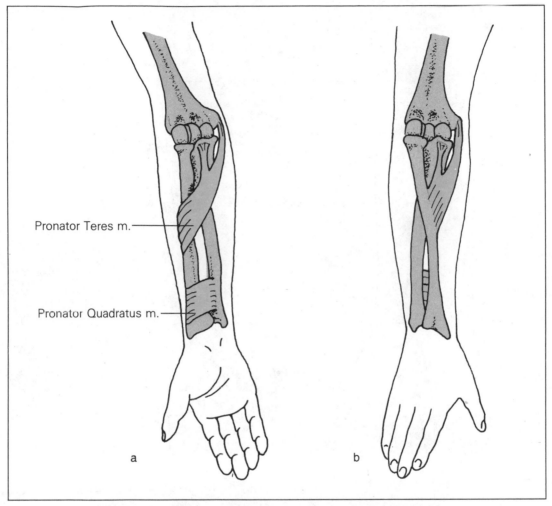

Fig 3–54.—The pronators of the forearm: **(a)** before and **(b)** after pronation.

and (2) the supinator m., which is also an extensor of the elbow joint, is slightly prestretched during flexion and thus can exert its greatest power in this position.

The Articulations of the Wrist

The function of the wrist must be viewed from the same principal perspective as the functions of the shoulder and elbow joints, namely, the extension and differentiation of the reach, touch, and expression functions of the hands. The movements of the wrist are independent of the movements of the fingers and contribute to the increased usefulness of the hand.

The Proximal and Distal Wrist Joints

We distinguish a proximal and distal wrist joint (see Fig 3–58).

The *proximal wrist joint* is ovoid (ellipsoid), which allows both a palmar and dorsiflexion and also, a radial and ulnar abduc-

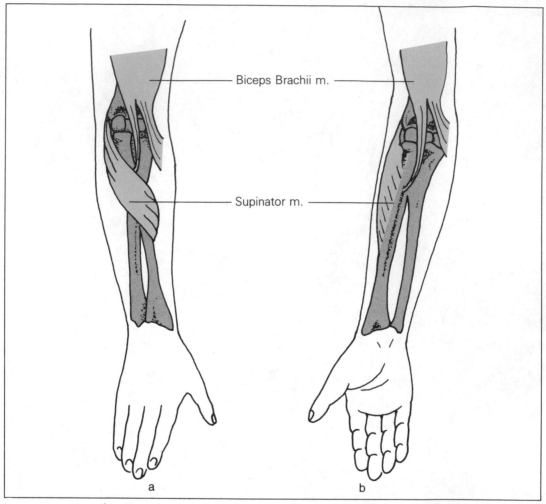

Fig 3–55.—The supinator m. and the biceps brachii m. **(a)** before and **(b)** after supination.

tion. The joint is composed of the radius, the articular disk of the ulna, and the proximal row of the carpal bones.

In the *distal wrist joint,* the proximal and distal row of carpal bones form an S-shaped articulating surface.

Functionally, the *proximal* and *distal wrist joints* act in unison, although palmar flexion occurs primarily in the proximal joint, while dorsiflexion occurs primarily in the distal wrist joint. In the abduction motions, the reciprocal tilting and sliding mo-

tions of the carpal bones allow an ulnar deviation of 40° and a radial deviation of 15°.

The Muscles Acting on the Wrist Joints

The muscles of the wrist ulnar and radial flexors and extensors do not insert in the wrist, as their Latin name would imply *(flexores et extensores carpi ulnaris et radialis),* but in the middle part of the hand. The labile functional relationship of the carpal bones would not permit the effective insertion of muscles.

The Flexors of the Wrist Joint

Flexor Carpi Ulnaris m. (Fig 3–56)

Origin: One head originates from the medial epicondyle of the humerus, the other from the olecranon

Insertion: The base of the fifth metacarpal (little finger). The tendon surrounds the fusiform bone that thus functions as a sesamoid bone.

Innervation: The ulnar n.

Function: In cooperation with the *extensor carpi ulnaris m.,* it abducts the wrist toward the ulnar side; in cooperation with the other flexors, it affects palmar flexion (see below). By virtue of its origin on the ulna, it also assists in flexing the elbow joint.

Flexor Carpi Radialis m. (Fig 3–56)

Origin: Median condyle of the humerus
Insertion: The base of metacarpal II

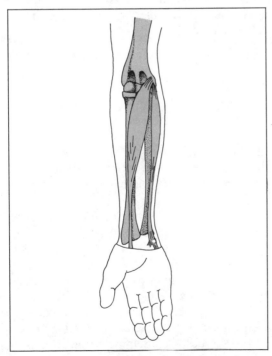

Fig 3–56.—The flexor carpi ulnaris m. and the flexor carpi radialis m.

Innervations: The median n.

Function: In cooperation with the *extensor carpi radialis m.,* it abducts the wrist radially, as, for example, in the final finger jerk at the moment of releasing the discus in a discus throw. It flexes the hand in cooperation with the *flexor carpi ulnaris m.* and the long flexors of the fingers (see p. 96). In addition, it assists in pronation and in flexing the elbow joint.

The flexors of the wrist and the superficial and deep flexors of the fingers (see p. 96), play an important role in all athletic activities in which a strong wrist or finger action is required. These include shot put, gymnastics (particularly in vaulting and summersaulting), climbing, etc. Figure 3–61 gives an overview of the superficial muscles of the palmar surface. It can be seen that the flexors of the forearm all insert in the region of the medial condyle of the humerus.

Extensor Carpi Ulnaris m. (Fig 3–57)

Origin: Lateral condyle of the humerus
Insertion: The base of metacarpal V
Innervation: Radial n.

Function: In cooperation with the already mentioned *flexor carpi ulnaris m.,* it causes the ulnar abduction of the wrist. It also extends the wrist dorsally, in cooperation with the extensors of the wrist and fingers to be discussed below.

Extensores Carpi Radialis Longus and Brevis m. (see Fig 3–57)

Origin: Lateral condyle of the humerus
Insertion: The base of metacarpal II and III
Innervation: Radial n.

Function: In cooperation with the *flexor carpi radialis m.,* they cause radial abduction. In cooperation with the *extensor carpi ulnaris m.* and the long extensors of the fingers, they produce the dorsal extension of the hand. One portion of the extensor carpi

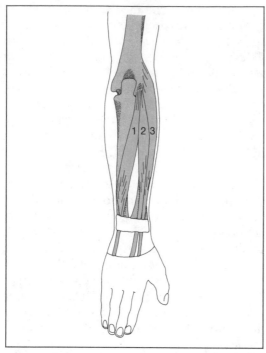

Fig 3–57.—The extensor carpi ulnaris m. **(1)**, the extensor carpi radialis brevis m. **(2)**, and longus m. **(3)**.

radialis longus m., which originates from the lateral intermuscular septum anterior to the rotational axis of the elbow joint, assists in the flexion of the elbow joint.

In the range of athletic movements, the extensors of the hand play only a subordinate role. Only in weight lifting and in fencing are they used to any great extent. They are much less developed than the corresponding flexors.

All extensors of the wrist originate on the *lateral* condyle of the humerus. Since the strain of a backhand stroke in tennis falls almost entirely on the extensors, it is not surprising that the lateral condyle is the focus of all symptoms in the ''tennis elbow.''

The versatility of the muscles of the wrist is explained by their location on the edges, between the flexors on the palmar surface of the forearm and the extensors on the dorsal aspect. The muscles of the wrist participate in all movements of the wrist joints. A simultaneous isometric contraction of both the flexor and extensor muscle groups results in a fixation of the wrist. This is necessary in performing handstands, straight arm blows, and all types of grappling holds in wrestling.

The Hand

The characteristic structure of the hand is related to its function as a grasping tool. This grasping ability is made possible by the fact that the thumb can be opposed to the fingers. The fingers and the thumb act together as a versatile and multipurpose pair of pliers. They need the palm of the hand as flat base, on which the object grasped and held can rest. To accomplish these functions, the hand is composed of three parts: the carpal area (wrist), the metacarpal area (middle part of the hand), and fingers (phalanges or digits).

The Osseous and Ligamentous Structure

Carpal and Metacarpal Area

As shown in Figure 3–58, the carpal area is composed of two rows of carpal bones. The proximal row consists of the scaphoid, the lunate, the triangular and the pisiform bones. The distal row consists of the trapezius, the trapezoid, the capitate, and the hamate bones.

The carpal bones have cartilaginous articulating surfaces at the points of contact and are so arranged as to form an arch. In this, they are similar to the metacarpal bones.

This arch is held together by a complex system of ligaments. It is bridged by a transverse ligament (transverse carpal ligament) under which the flexor tendons of the fingers proceed to their insertion in a common tendon sheath. A similar ligament fixes the extensor tendons to the carpal bones at the dorsum of the hand.

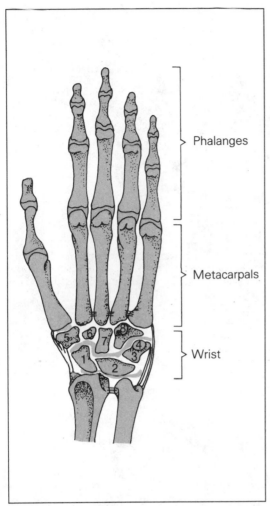

> Phalanges

> Metacarpals

> Wrist

Fig 3–58.—Representation of the wrist, metacarpals and phalanges (palmar view). Red lines indicate the extent of the proximal and distal wrist joints and the saddle joint of the thumb. (For the name of the carpal bones, see text.)

The distal row of carpal bones is attached to the metacarpal bones by such tight ligaments that the articulations here are transformed into synarthroses. The only exception is the saddle joint of the thumb.

The Thumb and Finger Joints

THE ARTICULATIONS OF THE THUMB.—The manifold motions of the thumb are made possible by the saddle joint of the thumb.

This joint is formed by the trapezoid and the base of the first metacarpal bone. This joint has two degrees of freedom (e.g., is biaxial) and allows for opposition, abduction, adduction, and circumduction of the thumb.

Opposition is *the* motion of the thumb, which allows the hand to function as a grasping tool. A complicated system of muscles allows the thumb to touch all fingers and some parts of the palm, and makes a variety of grasping and holding functions possible. The eight intrinsic muscles of the thumb will not be discussed in detail in this volume.

The thumb also has a proximal and a distal hinge joint. It lacks the middle hinge joint of the fingers.

THE ARTICULATIONS OF THE FINGERS.—The fingers have a proximal, middle, and distal joint.

The *middle* and *distal* articulations are hinge joints. The *proximal* joint is anatomically a ball and socket joint, but the tight ligaments limit it functionally to only biaxial movements: flexion and extension, abduction and adduction.

The special structure of the proximal finger joint gives an important characteristic to the grasping function of the hand: when flexed, it increases the stability of the joint, when extended, the joint is much looser. Such a mechanism is the prerequisite for a solid grasp. It is accomplished by strong collateral ligaments, which are so arranged that they are tightened by flexion and loosened by extension (Fig 3–59).

THE MUSCLES OF THE FINGERS.—Just like the thumb, the fingers also have their intrinsic muscles. Flexion of the basal, middle, and distal finger joint is accomplished by individual muscles, which allow individual positioning of the fingers. Extension is accomplished by a common dorsal aponeurosis to which the extensor muscles are attached.

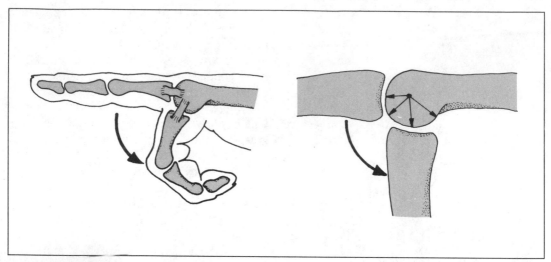

Fig 3–59.—The structure of the phalangeal joints and the function of the collateral ligaments. During flexion, the collateral ligaments become tighter, since the distance between the distal part and the axis of rotation is increased (after *Rohen*).

To avoid a troublesome thickening of hand, which would interfere with its grasping function, the bellies of the finger muscles are located in the forearm and not in the hand.

The distal joints of the fingers are flexed by the *flexor digitorum profundus m*. The middle joints are flexed by the *flexor digitorum superficialis m*. (Fig 3–60). The tendons of the superficial flexors are divided at their insertion, and through this slit, the tendons of the deep flexors pass to insert at the distal phalanx.

Extension of the fingers is accomplished by the *extensor digitorum communis m*.

Only the function of these muscles will be discussed. Their details and the description of the small, intrinsic muscles of the fingers is omitted.

For maximal precision of the range of motion of the fingers, there are *dorsal and palmar interosseous muscles* and *lumbricoid muscles* (Fig 3–61). These originate on the flexor surface of the basal joint, and insert on the extensor suface of the middle and distal phalanges. This allows for a staggered

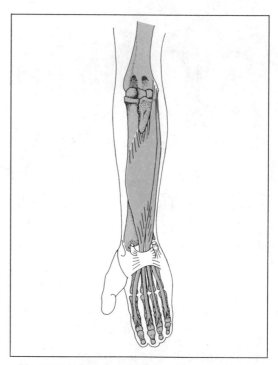

Fig 3–60.—The flexor digitorum superficialis m.

progression of the flexion of the fingers. If the fingers were to have only the long flexors, every flexion of the basal joint would also lead to a flexion of the middle and distal joints. This would make the ability to grasp very much cruder and less precise.

The ability of these intrinsic muscles to affect abduction and adduction (spreading) of the fingers adds a further dimension to the mobility of the hand.

The Lower Extremity

In principle, the structure of the lower extremity is similar to the upper extremity. The erect posture and the consequent static load on the legs has resulted in some structural peculiarities in the lower extremity, in response to these functional demands.

The pelvic girdle, in contrast to the highly mobile shoulder girdle, forms a rigid, closed ring (see p. 55), which supports the trunk and which also serves as an abutment for the lower extremities and a base for their mobility.

In contrast to the upper extremity, the articulations become distally more and more restricted to provide increased stability and security.

Pronation and supination are present only in a reduced form and are limited to the foot. The two bones of the leg cannot move in relation to each other.

The proximal ankle joint, which corresponds to the proximal carpal joint, allows only for flexion and extension, in the service of forward motion.

The Hip Joint

Osseous and Ligamentous System

The hip joint consists of the head of the femur and of the acetabulum of the hip bone. While the shoulder joint is controlled by muscles only, the hip joint, a true ball and socket joint, is controlled by bone, tendon, and muscle.

The *osseous control* is achieved by placing the head of the femur deeply into the acetabulum. In addition, the acetabulum is extended by a cartilaginous fiber lip, which increases the articulating surface between the bones and further strengthens the joint.

The *ligamentous control* is achieved by an extraordinarily strong group of ligaments, which are arranged in a spiral fashion. When the hip joint is extended, the ligamentous spiral is further tightened. A "lateral split" is therefore easier to perform when the hip is flexed than when it is fully extended.

The four ligaments, which participate in the formation of this ligamentous spiral, are the iliofemoral, the pubofemoral, the ischiofemoral and the zona orbicularis. Of these, the iliofemoral is the most important. It is the *strongest ligament in the body* and has a tensile strength of 300 kg. It consists of two V-shaped components of which the vertical one limits the retroversion of the leg and the horizontal one limits the adduction of the leg (Fig 3–62). The primary purpose of the ligaments of the hip, actively supported by the muscles of the hip, is not so much to limit the motions of the leg but rather to assure the position of the pelvis and the trunk. The vertical strand of the iliofemoral ligament prevents the trunk from tipping backward in the standing position. The horizontal strand keeps the upper body from tipping toward the moving leg, in walking, at the time when the pelvis has to be tilted from standing leg toward the moving leg and the body has to be balanced on the femoral head of the standing leg.

The shaft of the femur is slightly tilted away from the main weight-bearing axis and forms an *angle* with the femoral neck (see Fig 3–62), which is 150° in the new-

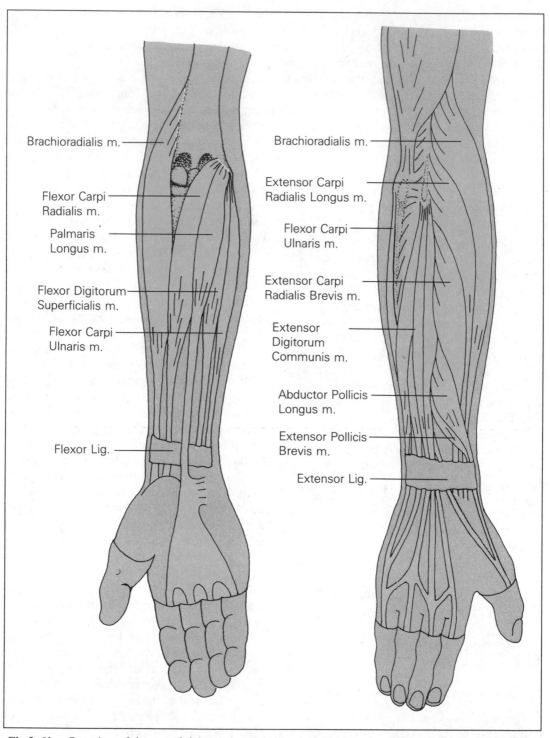

Brachioradialis m.

Flexor Carpi
Radialis m.

Palmaris
Longus m.

Flexor Digitorum
Superficialis m.

Flexor Carpi
Ulnaris m.

Flexor Lig.

Brachioradialis m.

Extensor Carpi
Radialis Longus m.

Flexor Carpi
Ulnaris m.

Extensor Carpi
Radialis Brevis m.

Extensor
Digitorum
Communis m.

Abductor Pollicis
Longus m.

Extensor Pollicis
Brevis m.

Extensor Lig.

Fig 3–61.—Overview of the superficial muscles of the forearm. Palmar view **(left)** and dorsal view **(right).**

Mossinger in a superpower
maneuver on the still rings.

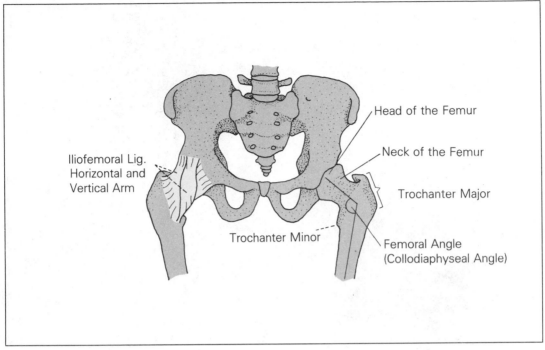

Fig 3–62.—The osseous and ligamentous system of the hip joint.

born, but only 120° in the adult. This angulation provides good leverage to the hip muscles, which insert in this area.

As previously discussed (see p. 55 and Fig 3–9) the pelvis is able to carry the weight of the trunk by virtue of its vaultlike construction. A similar construction exists between the pelvis and the femur, where the tension lines go through the pubic symphysis. If the symphysis separates (overload or trauma), the static equilibrium of the pelvis is destroyed, resulting in a significant limitation in walking.

The Muscles Acting on the Hip Joint

The range of motion of the leg is largely limited to the field of vision. All motions of the lower extremity that are outside the field of vision are limited, in order that the support structure of the trunk and its base, the foot, never strays far from the control field of the eyes.

The Ventral Muscles of the Hip

Iliopsoas m. (Fig 3–63)

This muscle is composed of two distinct parts that have a particular site of origin. They are the *psoas m.* and the *iliac m.*

Origin: The *psoas m.* originates on the last thoracic vertebra, on the first four lumbar vertebrae and on the costal processes. The *iliac m.* originates from the iliac wing and from the anterior, inferior iliac spine.

Insertion: The lesser trochanter of the femur

Innervation: Lumbar plexus

Function: On the swinging leg side, this muscle anteflexes, externally rotates, and adducts. On the side of the standing leg, it tilts the trunk laterally and anteriorly.

In addition, the *iliopsoas muscle* has an important role in stabilizing the pelvis. It tilts the pelvis forward, and cooperates with the other flexors of the hip in opposing the abdominal and gluteal muscles (see below).

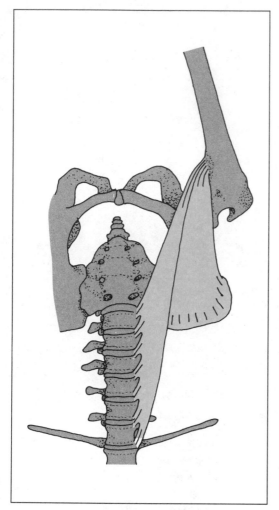

Fig 3–63.—The iliopsoas m.

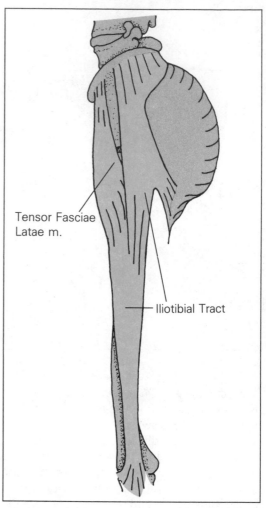

Fig 3–64.—The tensor fasciae latae m. and the iliotibial tract.

This muscle is a *typical running muscle,* which pulls the thigh anteriorly. Its strength and endurance determines the length of the stride and the consistency of the stride in, for example, the 400-m dash. It is, simultaneously, one of the most important muscles in the straight, forward kick in soccer. In gymnastics, it also plays an important role in all exercises in which the legs have to be brought forward from the straight vertical position (e.g., the ''L'' support on the parallel bars).

Tensor Fasciae Latae m. (Fig 3–64)

Origin: Anterior, superior iliac spine

Insertion: The iliotibial tract and the lateral condyle of the tibia

Innervation: The superior gluteal n.

Function: It guides the thigh of the swinging leg foward and abducts it. On the standing leg side, it tilts the trunk and pelvis forward.

In addition, the *tensor fasciae latae m.* plays an important role in tightening the fascia lata, which serves as the insertion site for the *gluteus maximus m.* (see p. 106). This helps to stabilize the thigh. As mentioned above, the femur is subject to bending stresses because of its angulation in relation to the weight-bearing axis of the leg. This stress can be counteracted by a pull on the fascia lata, which is attached to the iliotibial tract. The importance of this mechanism can be illustrated by the following example: in every jump from varying heights, different bending forces are exerted on the thigh. It is, therefore, essential that the anticipated stress, which depends on the height of the jump, be opposed by a corresponding, preprogrammed counter stress. This is possible only if there is an adaptable muscular system. A tight, rigid fascial system could not cope with the variable stress, and there would be an increased risk of fractures. This is the case when the height of fall is misjudged, as, for example, when missing a step on the staircase.

Rectus Femoris m. (see Figs 3–75 and 3–78)

The *rectus femoris m.* is a component of the *quadriceps femoris m.* (for details see p. 110), which, being biarticular, acts as a flexor of the hip joint, tilts the pelvis forward, and assists in stabilizing the pelvis.

Sartorius m. (see Figs 3–77 and 3–78)

This muscle, which is also biarticular, flexes the hip joint, rotates it externally, and abducts it. (The function of the sartorius m. on the knee joint is discussed on p. 113.)

The Adductor Group

The adductor group lies on the internal side of the thigh and is wedged in between the flexor and extensor groups. The adductors consist of three layers.

The *superficial layer* (Fig 3–65) is composed of the *pectineus m.,* the *adductor longus m.,* and the *gracilis m.* This last one is the only biarticular muscle of the group.

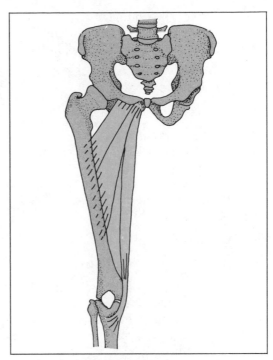

Fig 3–65.—The superficial layer of the adductors of the thigh: the pectineus m. (**above**), the adductor longus m. (**middle**), and the gracilis m. (**below**).

Pectineus m. (see Fig 3–65)

Origin: The pecten of the pubis
Insertion: The pectineal line of the femur
Innervation: The femoral and obturator nerves
Function: This muscle adducts the thigh and assists in flexing and externally rotating the hip joint.

Adductor Longus m. (see Fig 3–65)

Origin: The pubic tubercle
Insertion: The middle third of the medial lip of the linea aspera
Innervation: The obturator n.
Function: This muscle adducts the thigh and assists in flexing the hip joint.

Gracilis m. (see Fig 3–65)

Origin: Inferior pubic ramus
Insertion: Medial edge of the tibial tuberosity as part of the pes anserinus

Innervation: The obturator n.

Function: This biarticular muscle flexes the hip joint and also flexes and internally rotates the knee joint

The Middle Layer

Adductor Brevis m. (Fig 3–66)

Origin: Inferior pubic ramus

Insertion: Proximal third of the linea aspera

Innervation: The obturator n.

Function: This muscle flexes and externally rotates the thigh

The Deep Layer

Adductor Magnus m. (Fig 3–67)

Origin: The ramus of the ischium and the inferior edge of the ischial tuberosity

Insertion: One part inserts in the medial lip of the linea aspera, the other on the medial femoral epicondyle.

Innervation: The obturator and tibial nerves.

Function: *The adductor magnus m.* is the most powerful adductor muscle of the thigh which, by virtue of its lower insertion, also has an internal rotatory effect. This internal rotatory component is due to the fact that one part of this muscle originates from the ischial tuberosity and hence is posterior to the rotational axis.

A summary discussion of these five adductor muscles must emphasize their dynamic and *static* effects. The *primary dynamic function* of the adductors is to adduct the abducted legs (hence the name!). Of additional importance is the fact that the adductors can also function as strong flexors or extensors of the hip joint, dependent on their location anterior or posterior to the flexion-extension axis of the hip joint.

In walking and running, the contraction of the adductors contributes to the forward and backward motion of the swinging leg.

The *primary static effect* of the adductors consist in stabilizing the labile equilibrium

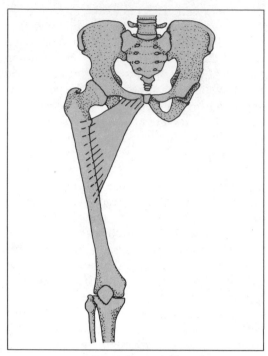

Fig 3–66.—The adductor brevis m.

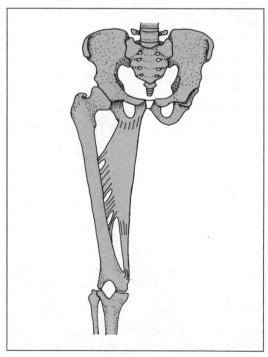

Fig 3–67.—The adductor magnus m.

of the trunk by a constant adjustment of the position of the pelvis. The adducting, internal, and external rotating components of these muscles prevents the twisting of the pelvis.

The Lateral Muscles of the Hip

The antagonists of the adductors, *the abductors,* can be found on the lateral surface of the pelvis. This muscle group, which is covered by the gluteus maximus m. (see p. 106), is of extraordinary importance in normal locomotion.

Gluteus Medius m. (Fig 3–68)

Origin: The external surface of the ilium
Insertion: The greater trochanter
Innervation: The superior gluteal n.
Function: The most important function of this muscle is the abduction of the thigh (e.g., the lateral split on the horizontal bar). When the thigh is fixed (standing leg), unilateral stimulation will lead to a lateral flexion of the trunk. In walking and running, it prevents the pelvic tilt from the side of the support leg to the side of the swinging leg and thus contributes to the maintenance of the erect posture. Paralysis of this muscle interferes with normal locomotion to a significant degree and leads to the development of a "waddling" walk.

Besides this primary function, the *gluteus medius m.* also has an internal rotating component, by virtue of its ventral portion, and an external rotatory component, by virtue of its dorsal portion. Furthermore, its anterior portion participates in forward motion and its posterior portion in backward motion. When all portions contract simultaneously, the result will be abduction.

Gluteus Minimus m. (Fig 3–69)

Origin: This muscle, located under the *gluteus medius m.,* originates from the external surface of the ilium.
Insertion: The greater trochanter
Innervation: The superior gluteal n.

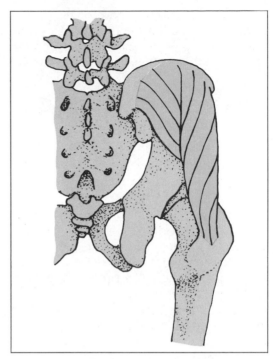

Fig 3–68.—The gluteus medius m.

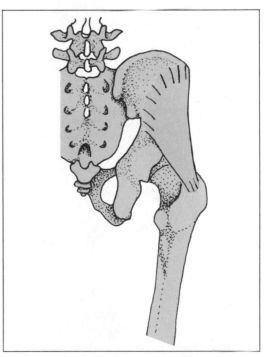

Fig 3–69.—The gluteus minimus m.

Function: Similar to those of the *gluteus medius m.* It abducts the thigh, rotates it internally, and pulls it forward. It can do this because part of the muscle is anterior to the rotational axis.

The Dorsal Hip Muscles

Gluteus Maximus m. (Fig 3–70)

Origin: The ilium, sacrum, coccyx, and the sacrotuberous ligament

Insertion: The fascia lata and the gluteal tuberosity of the femur

Innervation: The gluteus inferior n.

Function: This is one of the strongest muscles in the body. Its primary function is the extension of the hip joint as, for instance, in rising from a squatting position, in running, and jumping.

In addition, its upper portion functions as an abductor, its lower portion as an adductor. Finally, the muscle also has a powerful external rotatory function.

In addition to these dynamic functions, the *gluteus maximus m.* also has an important *static function.* By a tendinous insertion on the fascia lata, it participates in the splinting of the thigh (see p. 101), and also prevents excessive forward tilting of the trunk as, for example, in downhill skiing or speed skating. Furthermore, the *gluteus maximus m.* is very important in stabilizing the pelvis. It cooperates with the rectus abdominis m. in tilting the pelvis backward. If the gluteal muscles become weak, this leads to an exaggeration of the lumbar lordosis and the development of a ''sway back'' posture.

The Ischiocrural Muscles

This biarticular muscle group is discussed in some detail as part of the muscles acting on the knee joint (see p. 114). It assists in

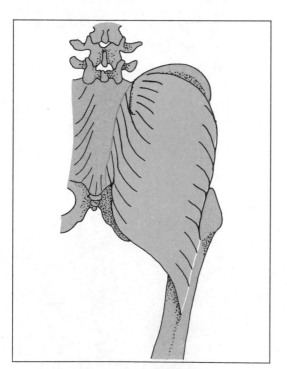

Fig 3–70.—The gluteus maximus m.

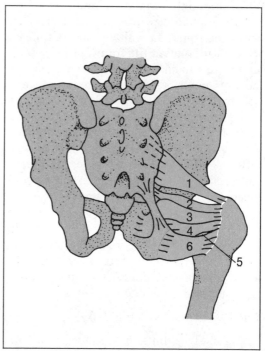

Fig 3–71.—The external rotator group of the thigh: **1,** the piriformis m., **2,** the gemellus superior m., **3,** the gemellus inferior m., **4,** the obturator externus m., and **5,** the quadratus femoris m.

the extension of hip joint and thus is synergistic with the *gluteus maximus m.*

Located under the *gluteus maximus m.* is a group of *external rotators.* They originate on the pelvis and proceed to the trochanteric fossa and the intertrochanteric crest. They will not be further discussed in this volume.

The following muscles are included in this group (Fig 3–71):

- Piriformis m.
- Gemellus superior m.
- Obturator internus m.
- Gemellus inferior m.
- Obturator externus m.
- Quadratus femoris m.
- Shared functions: as their group designation indicates, their primary function is external rotation. The *piriformis m.* also abducts, while all the other adduct. When the lower extremity is fixed, they tilt the pelvis laterally and flex it dorsally.

The Knee Joint

The Osseous and Ligamentous System

The articulating surfaces of the knee joint are the condyles of the femur and the medial and lateral condyle of the tibia. To provide stability to the leg as a "support column," in contrast to the elbow, only two, rather than three bones, participate in forming the joint.

The knee is a *hinge joint (gynglimus)* with two degrees of freedom. It allows flexion and extension and also some rotation, the latter only in the flexed position.

The knee joint connects the leg-foot lever with the thigh. It is the mobility of the latter that determines the field of motion of the former, which for reasons of static safety, is much more restricted than that of the hand. The functional unity of the thigh and leg, as a support column, is assured through the knee in the extended position.

In the flexed position, rotation of the knee allows an increased range of motion for the foot.

To overcome the differences in contour and to smooth out the sharp contact points between the articulating surfaces of the femur and tibia, the knee joint contains two articular discs, the lateral and medial *menisci.* These menisci, together with a system of ligaments and with the patella, make the knee joint into a very stable, but complicated joint (Fig 3–72). The *menisci* not only level the articulating surfaces, but also protect these surfaces against mechanical trauma, which may be caused by sudden increases in pressure.

The menisci adhere firmly to the condyles of the tibia, but are not anatomically a part of the cartilage of this bone. They are connected to the intercondylar ridge. The outer edges of the *menisci* are thick but taper sharply toward the central fossa. The *medial meniscus* is weaker than the lateral one, and just barely forms a semicircle. The *lateral meniscus* is an almost complete circle, which is open only where it is attached to the intercondylar ridge.

During movement of the joint, the *men-*

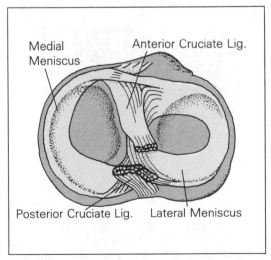

Fig 3–72.—The medial and lateral meniscus (from below).

isci also move, to give the condyles the widest possible support surface.

Digression: Meniscus Injuries

The medial meniscus is 20 times more often injured than the lateral meniscus. This is due to the fact that the medial meniscus, being attached to the joint capsule and to the median collateral ligament, cannot escape as readily from external, traumatic impact.

The most common situation for meniscus injury is the one where the flexed, abducted joint is suddenly extended, while the leg is externally rotated and the foot is fixed.

Clinical hint: surgical removal of the menisci leads, in about 15 years, to a fusion of the knee joint.

The two *cruciate ligaments* (Fig 3–73) are the restraining bands of the articulating condyles. In the weak, flexed position, when the collateral ligaments are relaxed,

the anterior displacement of the tibial head is prevented by the anterior cruciate ligament, and posterior displacement is prevented by the posterior cruciate ligament. In every position of the knee joint, some part of the cruciate ligaments is under tension.

During internal rotation of the leg, the two cruciate ligaments are twisted around each other and this acts as a brake for this type of motion. Since during external rotation the cruciate ligaments untwist, there is much less limitation on this motion. Rupture of the anterior cruciate ligament is usually accompanied, for reasons already discussed, by injury to the medial meniscus, the ligament, and the joint capsule.

The two *collateral ligaments* (Fig 3–74) protect the knee joint during extension. During flexion they are relaxed, but during extension they are maximally tightened.

The typical mechanism for rupture of the collateral ligaments is force applied to the

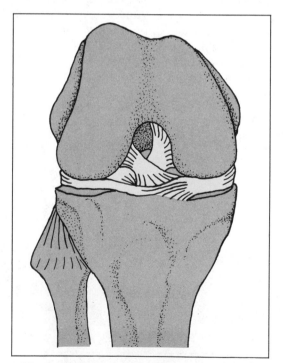

Fig 3–73.—Anterior and posterior cruciate ligament (with the patella removed).

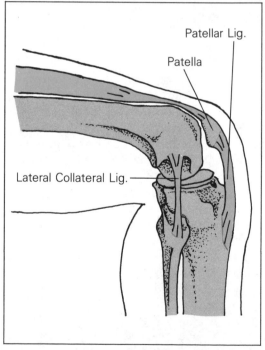

Fig 3–74.—Representation of the lateral collateral ligament, the patella, and the patellar ligaments.

extended knee from the side. This is what happens when, during a "sliding tackle," an offensive player falls onto the extended leg of a defensive player (soccer or American football).

All areas of the knee joint, which are subject to significant mechanical stresses, are supplied with bursae. The patella (see Fig 3–74) is another component that serves to improve the mechanical function of the knee joint. It is the largest sesamoid bone in the human body. It is embedded into the terminal tendon of the *quadriceps femoris m.*, and not only serves to align this tendon, but also improves the leverage of this important extensor muscle of the knee joint. The continuation of the quadriceps tendon beyond the patella is known as the patellar tendon (see Fig 3–74). The insertion of the muscle onto the tibial tuberosity is accomplished via this tendon.

The inner aspect of the patella is covered with cartilage, which reduces the friction of the articulating surfaces of the knee joint to a minimum. If this cartilage is excessively burdened by unphysiologic demands put on the knee joint, for example, extensive deep knee bend exercises under heavy loads and for extensive periods of time, a condition known as *chondropathia patellae* may ensue. This is a degenerative disease of the patellar cartilage that will lead to chronic knee joint complaints.

Note: In a half knee bend, the traction forces of the knee joint ligaments is increased by a factor of 5 to 6; in a deep knee bend by the factor of 12, and at this level they stand at 1,000 kp!

The Terminal Rotation of the Knee Joint

The terminal rotation of the knee joint is an additional measure to assure the stability of the straight leg. At the end of the regular extension, there is an additional extension of about 10°, which is made possible by an earlier external rotation of the leg (about 5°). This terminal "locking" of the straight leg, which makes all rotatory motions impossible, is accomplished by the anterior cruciate ligament. By full extension of the knee this ligament is tightened to the point where it rotates the tibia externally and the thigh internally. Flexion of the knee is possible only when this terminal extension had been reversed.

In summary, it can be said that the osseous and ligamentous systems of the knee joint serve the security of the leg as a "support column" and allow increased play only when the erect posture and walk is not in

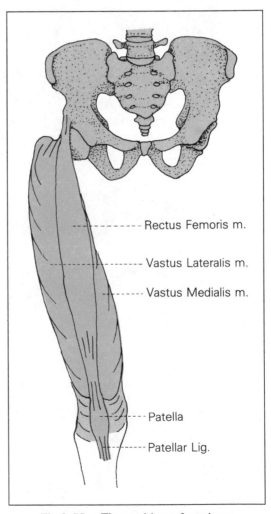

Fig 3–75.—The quadriceps femoris m.

Rectus Femoris m.

Vastus Lateralis m.

Vastus Medialis m.

Patella

Patellar Lig.

danger. This occurs only when the knee joint is flexed, i.e., carries no weight.

The Muscles Acting on the Knee Joint

Since the extensor muscles must support the weight of the entire body, while the flexor muscles carry only the weight of the leg, it is easy to see why the extensor muscles predominate in the lower extremity. Such a predominance of the extensors is an absolute requirement for the erect posture and for normal walk.

The Muscles of the Anterior Thigh

Quadriceps Femoris m. (Fig 3–75)

The *quadriceps femoris m.* is the principal extensor of the knee joint.

It is the largest and most powerful muscle in humans.

It is composed of the biarticular *rectus femoris m.* and the monoarticular *vastus medialis, lateralis, and intermedius m.*

Origin: The biarticular part of the muscle, the *rectus femoris m.*, originates on the anterior, inferior iliac spine and the upper rim of the acetabulum. The three other parts originate on the medial and lateral lip of the linea aspera and on the anterior and lateral surface of the femur.

Insertion: By way of the patellar ligament, on the tibial tuberosity

Innervation: The femoral n.

Function: The *quadriceps femoris* plays a towering role, both from a *dynamic* and *static* point of view. Its *static* role is to prevent the buckling of the knee while standing; its *dynamic* role is to forcefully extend the knee, as in all running and jumping exercises. The *rectus femoris m.*, in addition, also flexes the hip joint.

It is a peculiar characteristic of *the quadriceps femoris m.*, mandated by a functional necessity, that it is composed of two different muscle types. The *rectus femoris m.* consists of primarily FT fibers (see p. 195)

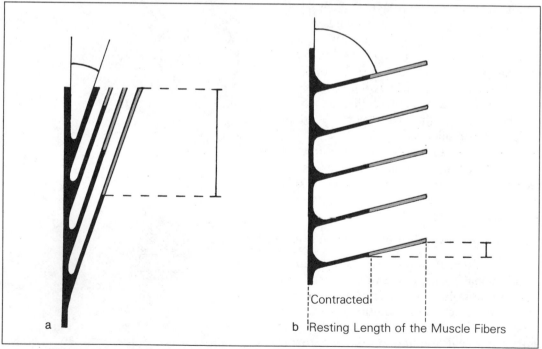

a

b Contracted

Resting Length of the Muscle Fibers

Fig 3–76.—Representation of the feathering angle and respective lifting distance of the rectus femoris m. and vastus muscles (contraction = black; resting state = red).

Wolfshohl on the way to
victory in the world
championship cross-country
race.

having an acute feathering angle. This is a "rapid action" muscle in which the degree of contraction is of great importance. In the other three components of the quadratus femoris m., the ST fibers are dominant, having an obtuse angle of feathering (see p. 195, and serving mostly for isometric holding actions, to assure the support column role of the lower extremity (Fig 3–76).

Tensor Fasciae Latae m. (see Fig 3–64)

This muscle, discussed earlier (see p. 102) plays a limited role in knee joint extension.

Sartorius m. (Fig 3–77 and 3–78)

This is the longest muscle in the human body. (Depending on the height of the individual, it may reach a length of 50 to 60 cm.)

Origin: Anterior, superior iliac spine

Insertion: Medial edge of tibial tubercle, on the pes anserinus (the pes anserinus is the common insertion of the sartorius m., *the gracilis m.,* and the *semitendinosus m.* (p. 115).

Innervation: The femoral n.

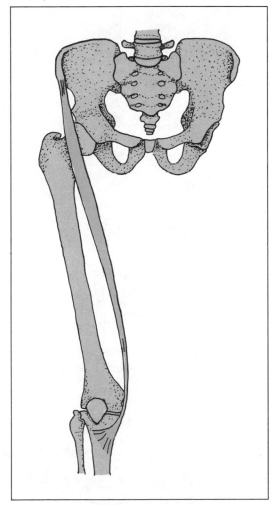

Fig 3–77.—The sartorius m.

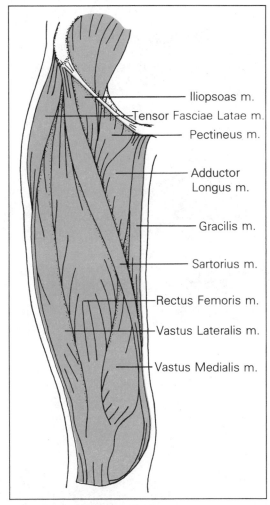

Iliopsoas m.
Tensor Fasciae Latae m.
Pectineus m.

Adductor Longus m.

Gracilis m.

Sartorius m.

Rectus Femoris m.

Vastus Lateralis m.

Vastus Medialis m.

Fig 3–78.—Overview of the muscles of anterior aspect of the thigh and hip.

Function: As a biarticular muscle, the *sartorius m.* assists in flexing, abducting, and externally rotating the thigh while also assisting in flexing the knee and internally rotating the flexed tibia.

This is the only muscle that flexes both the hip and the knee.

The Muscles on the Posterior Aspect of the Thigh

The muscles of the posterior aspect of the thigh, known as *ischiocrural m.,* have dy-

namic motion functions as their primary responsibility. In this, they differ from the knee joint extensors, the primary function of which is static, namely to provide stability to the system.

Figure 3–79 presents an overview of these muscles.

Biceps Femoris m. (Fig 3–80)

Origin: The *long head* from the ischial tuberosity, the *short head* from the lateral lip of the linea asperum

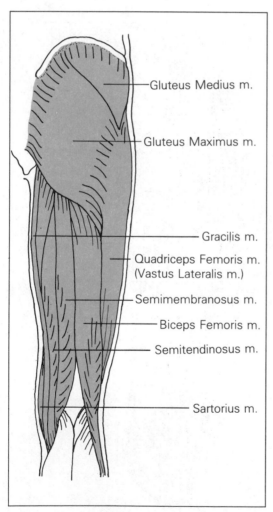

Fig 3–79.—Overview of the muscles of the posterior aspect of the thigh and hip.

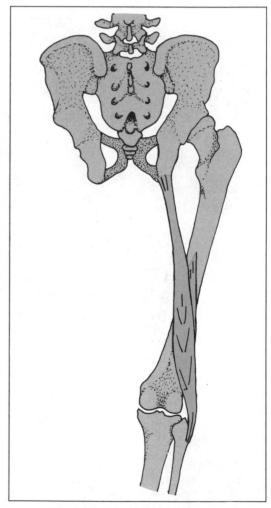

Fig 3–80.—The biceps femoris m.

Insertion: The head of the fibula

Innervation: The long head by the tibial n., the short head by the peroneal n.

Function: This biarticular muscle extends the hip (fixed leg), and also flexes the leg and rotates it externally in the flexed position (swing leg). In Alpine skiing, for example, this rotation of the leg is very important in pointing and controlling the skis. Other muscles, discussed below, participate in this activity.

Semitendinosus m. (Fig 3–81)

Origin: Ischial tuberosity

Insertion: Lateral to the tibial tuberosity in the pes anserinum

Innervation: The tibial n.

Function: This is also a biarticular muscle. Its function is very similar to biceps femoris except that it rotates the flexed leg internally.

Semimembranosus m. (see Fig 3–81)

Origin: Ischial tuberosity

Insertion: Medial condyle of the tibia

Innervation: The tibial n.

Function: The same as the *semitendinosus m.*, which covers it, but this muscle is more powerful.

The *ischiocrural m.* play an important role in walking, and are largely responsible for the vertical placement of the sole of the foot to the ground.

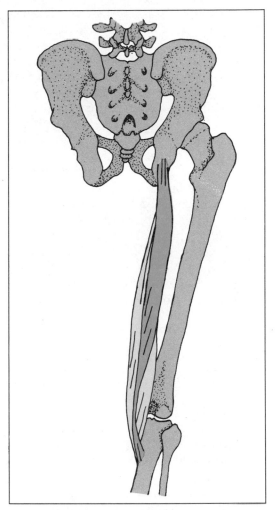

Fig 3–81.—The semitendinosus m. *(red)* and the semimembranosus m. *(pink).*

The Foot and the Articulations of the Foot

In describing the foot and the functional unity of the leg and foot, the similarity with the upper extremity comes forcefully to mind. In contrast to the upper extremity, the primary functions of the lower extremity are predominantly static. For this reason, the mobility of the individual joints decreases distally. The *upper ankle joint* is a pure hinge joint with one axis, which permits only flexion and extension. In comparison, the corresponding proximal carpal joint has two degrees of freedom. Rotations of a limited nature is granted to the foot only in the *lower ankle joint*. Pronation and supination, as compared with the upper extremity, are placed more distally for reasons of static security and are modified accordingly. The toes, which also predominantly serve static functions, are shorter and more robust than the fingers. The ankle bones are stronger than the carpal bones, which are designed for mobility, and are designed to support the lever action of the foot.

Contrary to the hand, which is a grasping tool and which is an extension of the forearm, the foot and leg constitute a hinge-lever system that serves for forward motion. The longer the heel, the better the hinge-lever action.

Similar to the pelvis, the foot forms a vault, which can elastically distribute the weight of the body. The musculature of the foot is also designed to serve a static purpose. Whereas in the upper extremity both flexors of the fingers are located in the forearm, to hinder the mobility of the hand as little as possible, in the lower extremity only the long flexor of the toes is located in the leg. The short flexor is located in the area of the sole and helps to support the arch of the foot.

The Structure of the Foot

The organization of the foot into tarsal, metatarsal, and phalangeal areas is similar to that of the hand. In contrast to the thumb, which has to be short to function well as an opposing digit, the big toe is the longest digit of the foot, and serves as the main support for the arch of the foot (Fig 3–82).

The largest tarsal bone is the calcaneus. Medially, it has a bony prominence, the sustentaculum, which serves as a contact point for the talus.

The distal tarsal bones are connected to the metatarsals through amphiarthroses. This allows only spinglike motion in this area. The bones of the foot are arranged so that they form *medial and lateral beams*.

The *medial* one consists of the first three toes, the three cuneiform bones, the navicular bone, and finally, the talus. The *lateral* one consists of the last two toes, the cuboid and the calcaneus. Since the two beams are located at an angle to each other, this allows for the formation of the *tarsal arch*. The *median beam* is raised up and touches the ground again only at the proxi-

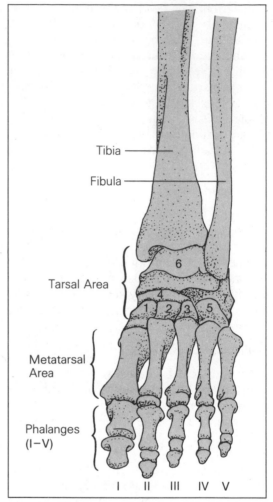

Fig 3–82.—Structure of the bony skeleton of the foot (top view). The ankle is composed of seven bones: **1,** the medial cuneiform, **2,** the intermediate cuneiform, **3,** the lateral cuneiform, **4,** the navicular, **5,** the cuboid, **6,** the talus and **7,** the calcaneus.

mal joint of the big toe. The *lateral* beam is in contact with the ground throughout its entire length. The position of the two beams and the organization of the plantar arch can be roughly equated, for the sake of simplicity, to two sticks placed at an angle across (over) each other (Fig 3–83).

As seen in the above illustration, the superimposition of the two beams produces not only a longitudinal, but also a transverse plantar arch. The transverse arch is the re-

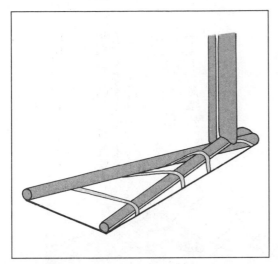

Fig 3–83.—Diagrammatic representation of the medial and lateral shaft of the foot, which form a longitudinal and transverse arch to distribute and equalize the weight.

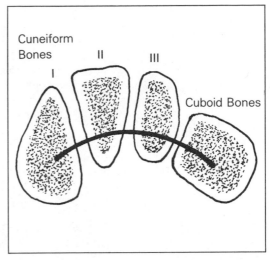

Fig 3–84.—Organization of the transverse arch of the foot, represented as a cross section at the level of the distal ankle bones.

sult of a particular arrangement of the three cuneiform bones and of the cuboid bone (Fig 3–84).

As shown in the following figure, the plantar arch is based on its osseous structure, but is also *actively* and *passively* braced and stabilized. This will be illustrated in a discussion of the longitudinal arch (Fig 3–85)

The *passive* bracing of the longitudinal arch is accomplished by three ligaments, the plantar aponeurosis superficially, the long plantar ligaments, and the deep calcaneonavicular ligament. The last one also participates in the distal plantar joint.

The *active bracing* of the longitudinal arch is accomplished by the muscles of the sole of the foot (see p. 120) and by the long muscles of the foot, active in the area of the lower leg. Their actions are a reflex response, regulated by pressure on the sole of the foot. For every change in tension on the arch, there will be a corresponding change in the tone and contraction of the muscles acting on the longitudinal arch. This fact also explains why it is more fatiguing to stand than to walk or to run, namely, stand-

ing causes a continuous contraction of the muscles and thus leads to a rapid fatigue. In occupations requiring standing, the arches may eventually flatten out. This process can be accelerated in those cases where there was already a preexisting weakness of the ligaments.

When the longitudinal arch is decreased, the resulting condition is known as "fallen arches," or "flat feet," which constitutes the most common of all orthopedic complaints. The decrease of this arch is the result of a medial displacement of the talus relative to the calcaneus. When the lateral arch is decreased, the resulting condition is known as "splay foot."

The Articulations of the Foot

In the area of the foot, we distinguish a superior and an inferior ankle joint.

The Superior Ankle Joint

The *superior* ankle joint is composed of the tibia, the fibula, and the trochlea of the talus. The medial and lateral malleoli of the tibia and fibula, respectively, envelop the trochlea like the jaws of a pair of pliers and

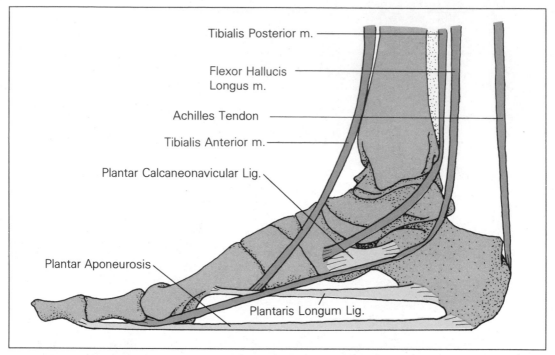

Tibialis Posterior m.

Flexor Hallucis
Longus m.

Achilles Tendon

Tibialis Anterior m.

Plantar Calcaneonavicular Lig.

Plantar Aponeurosis

Plantaris Longum Lig.

Fig 3–85.—Diagrammatic representation of the longitudinal arch of the foot.

form the so-called "malleolar fork." This hinge joint, which permits only flexion and extension, has a peculiar structural characteristic that rests on the configuration of the trochlea. The trochlea widens anteriorly and thus whenever the foot is flexed, the malleoli are being spread apart, become tighter, and finally lock the ankle, which stabilizes the foot. The closer the leg gets to the dorsum of the foot, the tighter the joint; the further they are from each other, the looser the joint (Fig 3–86).

The possibility of creating such a stable connection between the malleolar fork and the trochlea is totally dependent on the existence of an extraordinarily tough interosseous membrane between the tibia and fibula that guarantees that the malleolar fork does not spread.

A further security for the upper ankle joint is provided by a strong system of ligaments consisting of the interior and exterior collateral ligaments. The course of these ligaments is fan shaped, so that re-

gardless of the position of the joint, some part of the ligaments is tight and acts to stabilize the joint. This is particularly important since, anteriorly and posteriorly, this joint has only a loose articular capsule (Fig 3–87).

The Inferior Ankle Joint

The inferior ankle joint consists of two totally separate joints which, however, represent a functional unit. Since the talus articulates anteriorly with the navicular, but rests posteriorly on the calcaneus, there are, thus, two lower ankle joints, the *anterior* or talocalcaneonavicular joint, and the *posterior* or subtalar joint. The subtalar joint allows a rotatory movement around an oblique, anteromedial-posterolateral axis. Inward rotation of the foot (elevation of the external edge of the foot) is referred to as pronation. External rotation (elevation of the inner edge of the foot), is referred to as supination.

The supination and pronation of the foot

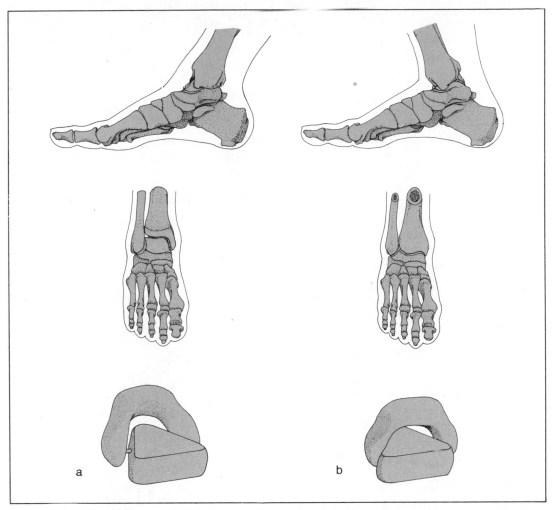

Fig 3–86. —The fixation mechanism of the proximal ankle joint with increasing approximation of the leg to the dorsum of the foot. In position **(a)** there is still a certain margin of play in the proximal ankle joint; in position **(b)** the lever action of the foot is fully engaged, i.e., it is fixated.

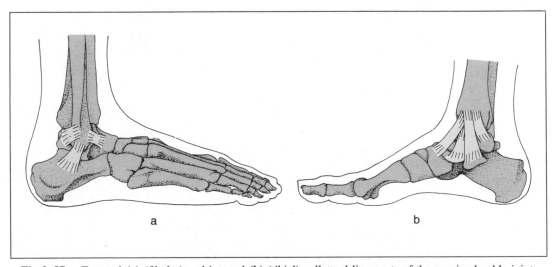

Fig 3–87.—External **(a)** (fibular) and internal **(b)** (tibial) collateral ligaments of the proximal ankle joint.

serve the purpose of adapting the foot to irregularities of the ground, assuring the widest possible area of contact and thus the maximum possible stability.

The anterior lower ankle joint has an additional peculiarity, which may become important in connection with a certain weakness of the foot. As shown in Figures 3–85 and 3–88, there is a space between the articulating surfaces of the navicular and the calcaneus, which is closed by the plantar, calcaneonavicular ligament. By congenital ligamentous weakness, or chronic excessive weight bearing, this ligament can become so loose that the head of the talus is displaced medially, the longitudinal arch disappears, and the condition known as "flat feet" is present.

The Muscles of the Foot

According to the functions of the foot, we distinguish two types of muscles: those that serve primarily a static purpose, such as the maintenance of the longitudinal and transverse arches in equalizing pressure, and those that serve primarily in forward motion.

The discussion of the short, intrinsic muscles of the foot, which correspond to the intrinsic muscles of the hand, although they serve a static rather than a dynamic function, will be omitted, and discussion will be limited to the muscles used in forward motion.

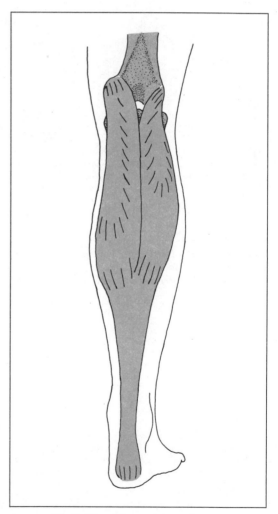

Fig 3–89.—The gastrocnemius m.

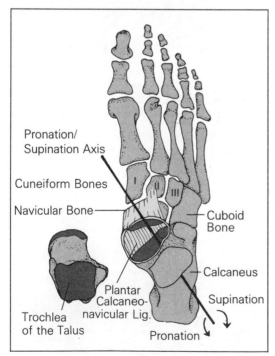

Fig 3–88.—Top view of the distal ankle joint (after removal of the talus) and representation of the supination-pronation axis of the distal ankle joint.

This attack scene at the
German Championship fencing
competition gives a good idea
of the extraordinary dynamics
of this activity.

The Muscles on the Posterior Aspect of the Leg

Since walking in the erect posture requires strong calf muscles to counterbalance the entire weight of the body, the flexor muscles of the leg, which bend the foot downward, are considerably more strongly developed than the extensor muscles. The extensor muscles form a single layer in the anterior aspect of the leg; the flexors consist of a superficial layer and a deep layer.

The Superficial Flexor Layer

Triceps Surae m.

This muscle consists of two distinct parts, i.e., the *gastrocnemius m.* and the *soleus m.*

Gastrocnemius m. (Fig 3–89)

Origin: Medial and lateral malleolus of the femur (the medial and lateral head)

Insertion: Via the achilles tendon on the posterior surface of the calcaneus

Innervation: The tibial n.

Function: Because of its primarily rapid, forceful action, this muscle is composed mostly of FT fibers (see p. 22). *The gastrocnemius m.* is significantly involved in plantar flexion of the foot and is thus instrumental in running and jumping. It lifts the heel off the ground and helps to ''push up'' at the ankle joint. In addition to its role as a supinator, this biarticular muscle also flexes the knee joint (swing leg).

The biarticular action of this muscle represents an important mechanism for many athletic activities, particularly jumping. The extension at the knee and the flexion at the ankle, so essential in walking, running, and jumping, prescribes that this muscle is always in an intermediate position. The plantar flexion in the accomplished sprinter or jumper occurs only when the knee is fully extended, and the muscle is markedly prestretched, and not when the knee is not fully extended and the contractility of the muscle is therefore reduced. The importance of this prestretching of the muscle becomes evident in one of the standard exercises used in training skiers. In this exercise, walking in a crouching position with the foot in plantar flexion is made particularly difficult, since the flexed position of the knee inhibits the prestretching of the *gastrocnemius m.*

Soleus m. (Fig 3–90)

Origin: The head of the fibula and the posterior surface of the tibia and fibula

Insertion: Via the achilles tendon on the calcaneus

Innervation: The tibial n.

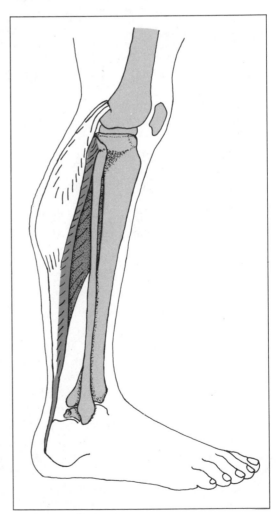

Fig 3–90.—The soleus m.

Function: In view of the identical insertion with the gastrocnemius m., this muscle is primarily a plantar flexor. Since it generates less contractile force than the *gastrocnemius m.,* it plays a lesser role in producing a maximal effort as, for example, in speed events, than in the endurance trials. Consequently, the muscle consists primarily of ST fibers (see p. 22).

The Deep Flexor Layer

The deep layer consists, from the medial to the lateral side, of the *flexor digitorum longus m.,* the *posterior tibial m.,* and the *flexor hallucis longus m.* (Fig 3–91).

Flexor Digitorum Longus m.

Origin: Posterior surface of the tibia
Insertion: Terminal phalanx of toes II through V
Innervation: The tibial n.
Function: This muscle serves both *dynamic* and *static* functions. It flexes toes II through V, assists in the plantar flexion of the ankle, assists in supinating the foot, and also supports the longitudinal arch.

Posterior Tibial m.

Origin: Posterior surface of the tibia and fibula and the interosseous membrane
Insertion: The navicular, the cuneiform bones, and the base of the first metatarsal bone
Innervation: The tibial n.
Function: The muscle supports plantar flexion in the upper ankle joint and supination in the lower ankle joint. It also supports the longitudinal arch of the foot.

The *posterior tibial m.* is particularly important in maintaining the longitudinal arch, since its tendon extends under the head of the talus and then spreads out in a fan-shaped structure to the inferior aspect of the navicular and medial cuneiform bones. It thus inserts on the highest point of the arch and prevents the ''falling'' of the arch and the medial displacement of the head of the talus.

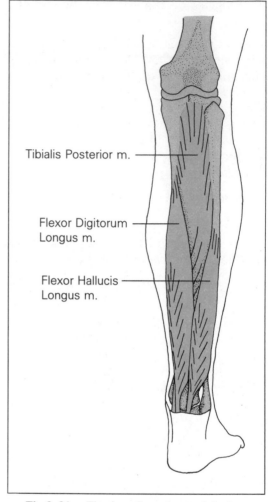

Tibialis Posterior m.

Flexor Digitorum Longus m.

Flexor Hallucis Longus m.

Fig 3–91.—The deep flexor layer of the leg.

Flexor Hallucis Longus m.

Origin: Fibula, interosseous membrane, and the intermuscular septum
Insertion: Terminal phalanx of the great toe
Innervation: The tibial n.
Function: This muscle participates in the plantar flexion of the ankle and flexes the big toe. It also supports the longitudinal arch.

The role of the *flexor hallucis longus m.* in supporting the longitudinal arch is particularly important, since, as already mentioned (see p. 117), this muscle extends under the sustentaculum of the talus and serves as an underpinning for it. It thus opposes the tendency of the calcaneus to buckle inward. This tendency of the calcaneus to buckle inward is due to the fact that the load lines of the foot impact on the talus and the calcaneus medially to the support point of longitudinal arch.

In summary, it can be stated that both the superficial and deep flexors participate to a greater or lesser extent in the plantar flexion of the upper ankle joint and in the supination of the lower ankle joints.

The Muscles of the Anterior Aspect of the Leg

Anterior Tibial m. (Fig 3–92)

Origin: Anterior surface of the tibia and the interosseous membrane

Insertion: Middle cuneiform and base of the first metatarsal

Innervation: The deep peroneal n.

Function: *The anterior tibial m.* dorsal flexes the foot, supinates it, and supports the arch. In combination with the *peroneus longus m.*, it forms the so-called stirrup (see p. 126). With the foot fixed, it pulls the leg forward and is thus used particularly in speed walking and skiing.

Extensor Digitorum Longus m. (see Fig 3–92)

Origin: Tibia, fibula, and interosseous membrane

Insertion: The dorsal aponeurosis of toes II through V

Innervation: Deep peroneal n.

Function: Dorsal extension of the foot and toes. In the lower ankle joint it supports pronation.

Extensor Hallucis Longus m. (Fig 3–93)

Origin: Fibula and interosseous membrane

Insertion: The dorsal aponeurosis of the great toe

Innervation: Deep peroneal n.

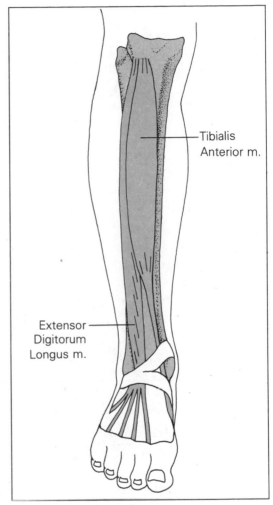

Tibialis Anterior m.

Extensor Digitorum Longus m.

Fig 3–92.—The tibialis anterior m. and extensor digitorum longus m.

Function: Dorsal extension of the foot and of the big toe. With the foot fixed, it assists in pulling the leg toward the foot and thus prepares for a jumping-off process. In this, it cooperates with the anterior tibial m. and the extensor hallucis longus m.

Note: Because of the lever action of the leg and the foot, and because of the 90° flexion inherent in this action, the muscles of the anterior compartment need some structure that prevents the tendons from separating from their base during contraction. This is accomplished by two ribbon-

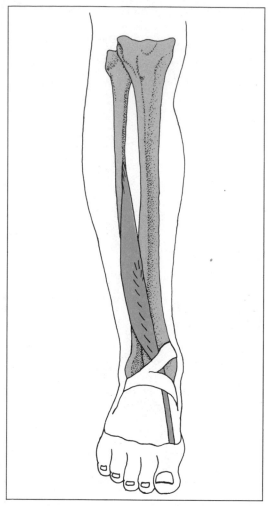

Fig 3–93.—The extensor hallucis longus m.

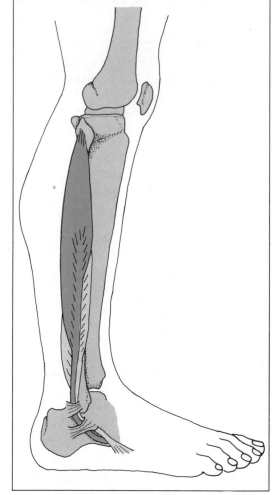

Fig 3–94.—The peroneus longus m. *(red)* and the peroneus brevis m. *(pink)*.

like structures, the superior and inferior retinaculum, located at the distal end of the leg and at the level of the ankle joint. These retinacula keep the tendons of the extensors in their proper position during dorsiflexion of the foot.

In summary, it can be stated that the muscles of the anterior compartment participate in dorsal extension at the upper ankle joint and in pronation at the lower ankle joint.

The Lateral Compartments of the Leg

Peroneus Longus m. (Fig 3–94)

Origin: External surface of the fibula, the intermuscular septa

Insertion: Middle cuneiform and tuberosity of the first metatarsal

Innervation: Superficial peroneal n.

Function: Plantar flexion at the upper ankle joint and pronation at the lower ankle joint. It forms the so-called stirrup with the *anterior tibial m.* The significance of this structure in maintaining the transverse and longitudinal arch of the foot is debatable.

Peroneus Brevis m.

Origin: External surface of the fibula and intermuscular septa

Insertion: Tuberosity of the fifth metacarpal

Innervation: Superficial peroneal n.

Function: Its dynamic function is the same as that of the *long peroneal m.*

4 Analysis of Simple Trunk and Extremity Movements

Introduction

The purpose of the ensuing discussion is to bring the reader to a rapid understanding of the anatomic principles of simple movements. The muscles that determine performance will be highlighted in a simplified, outline form, so that the muscles required to do the work can be strengthened through specific conditioning. It is in this context that the numerical values given should be interpreted. They are based on Lang and Wachsmuth and should serve to explain to the layman the individual contribution made by the muscles, which cooperate in a given movement. The author is aware that this approach tends to oversimplify the complexities of the locomotor process, but is willing to accept this handicap as a trade-off for the advantages of a simple and rapid overview.

The purpose of this section is then to bridge the gap between basic anatomy (which usually is a dry subject without any real relationship to reality) and the daily practice of athletics.

The Analysis of Simple Motion

Simple Trunk Movements

Forward Flexion

The amplitude of forward flexion of the trunk depends on the elasticity of the antagonists and of the ligaments of the spinal column. It also depends on the mobility of the small joints of the spinal column and on the power of the primary agonists.

When standing, gravity pulls the trunk forward until an equilibrium is established between the tone of the muscles responsible for the erect posture (mainly the *erector spinae muscles*) and gravity. When forward flexion continues, the same muscles, which assist in sit-ups from the supine position, are utilized.

Participating Muscles (Fig 4–1)

- *The rectus abdominis m.*
- *The external abdominal oblique m.*
- *The internal abdominal oblique m.*

And the hip flexors:

- *The rectus femoris m.*
- *The iliopsoas m.*
- *The tensor fasciae latae m.*
- *The sartorius m., etc.*

Note: When the knee and hip joints are flexed, the participation of the *rectus femoris m.* and *the iliopsoas m.* is markedly reduced, since their prestretch is decreased and their leverage is shortened. In this situation, the primary burden falls on the abdominal muscles.

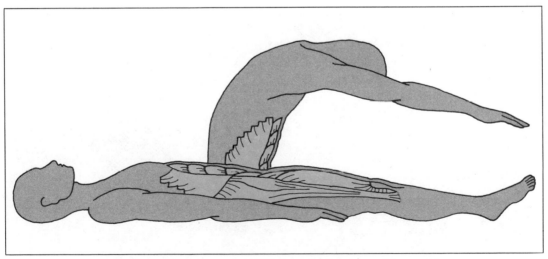

Fig 4–1.—The muscles participating in flexing the trunk forwards.

Dorsiflexion of the Trunk

In the dorsiflexion of the trunk, from the standing position, the degree of flexion is limited by the elasticity of the abdominal muscles. The elasticity of the ligaments of the spinal column and of the hip (mainly the iliofemoral ligament) and the mobility of the small joints of the spinal column are of great importance. The active participants in dorsiflexion of the trunk are the extensor muscles of the back.

Participating Muscles (Fig 4–2)

- *The erector spinae m.*
- *The latissimus dorsi m.* and the *trapezius m.* (they pull the shoulder backward)
- The *gluteus maximus m.* and the *ischiocrural m.* (they extend the hip joint)

Lateral Flexion of the Trunk

In the standing position, lateral flexion of the trunk is accomplished primarily by gravity, with the tone of the antagonists acting as brake. Only when an extreme flexion is required will the agonists come into play.

Participating Muscles (Fig 4–3)

On the ventral surface:
- The *rectus abdominis m.*
- The *external abdominal oblique m.*
- The *internal abdominal oblique m.*
- The *iliopsoas m.* (deep)
- The *pectoralis major m.* (it pulls the shoulder toward the contraction side)

On the dorsal surface:
- *The erector spinae m.* (in Fig 4–3,b these are covered by the *trapezius m.* and *latissimus dorsi m.*)
- *The quadratus lumborum m.*
- *The latissimus dorsi m.* and *the trapezius m.* (ascending portion). They, and several other muscles, pull the shoulder toward the contraction side.

The lateral flexion of the trunk is accomplished by the simultaneous contraction of the above muscles.

A Special Lateral Flexion of the Trunk.

—Raising the trunk from the horizontal, lateral position, with the legs fixed. To lift the pelvis in this maneuver, all muscles that originate on the body of the ilium and insert on the thigh or proximal leg must participate. A reversal of the fixed point

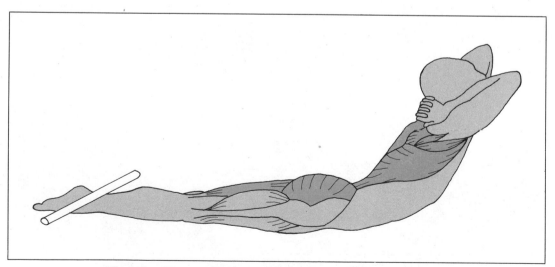

Fig 4–2.—The muscles participating in flexing the trunk backwards.

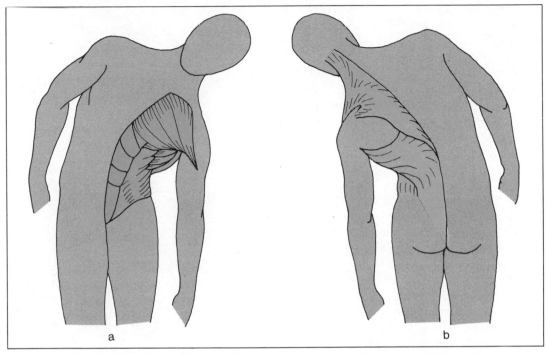

Fig 4–3.—The muscles participating in flexing the trunk sideways: **(a)** anterior view, **(b)** posterior view.

and mobile point of these muscles lifts the pelvis, rather than abducting the thigh, and thus supports the lateral raising of the trunk.

Participating Muscles

- *The gluteus maximus, medius, and minimus m.*
- *The rectus femoris m.*
- *The tensor fasciae latae m.*
- *The iliopsoas m.*

In gymnastics, a very similar combination of lateral flexion of the trunk and abduction of the hip is used in the scissor exercises in the pommel horse.

Rotation of the Trunk

Rotation of the trunk is accomplished by a serial contraction of the muscles having fibers oriented in a similar direction. This represents a form of muscle helix (Ben-ninghof) or muscle sling (Tittel) that rotates the trunk. On the dorsal surface, this muscle helix runs from the left side of the neck, over the left shoulder and left side of the trunk diagonally to the right anterior oblique muscle of the abdomen, and left interior oblique m. This is the sequence in levorotation of the trunk. In dextrorotation the sequence is the same, but in the opposite direction.

Participating Muscles

The participating muscles in this example of levorotation of the trunk are·

- Ventral aspect of the trunk:
- The right *external abdominal oblique m.*
- The left *internal abdominal oblique m.*

(since this muscle extends the direction of pull of the diagonally layered *external oblique* across the tendon sheath of the *rectus abdominis m.*)
- The *pectoralis major m.* (to pull the right shoulder forward)
- The *right serratus anterior m.* (functions like the pectoralis major m.)
- The *sternocleidomastoid m.* (to turn the head)

Dorsal aspect of the trunk: (Fig 4–4)
- The left *splenius m.* (to turn the head)
- The *transverse spinalis m.*
- The *levators* of the ribs
- The *external and internal intercostal muscles,* on the right and left side, respectively
- The right *external abdominal oblique m.* (see ventral aspect)

Additional contributions are made by some superficial muscles which, primarily, cause a rotation of the shoulder:
- Left part of the *latissimus dorsi m.*
- Left part of the *trapezius m.*
- Left *greater and lesser rhomboid m.*

Simple Movements of the Upper Extremity

In the following discussion of the extremity movements, the muscles will be described in a decreasing order of contractile power. The numbers of meter kilo pascal (mkp) that appear after the muscle, in parentheses, represent the calculated power output of the muscles from a resting position (Lanz and Wachsmuth). *Note:* these numbers serve only to assist in the evaluation of the muscles, which are important in that particular movement. They do not take into account the increase or decrease of muscle power, which depend on the relative lever positions, i.e., the varying angles of traction.

Anteversion (From the Anatomic Position)

This anteversion encompasses the movement from the dependent position to the anterior horizontal position.

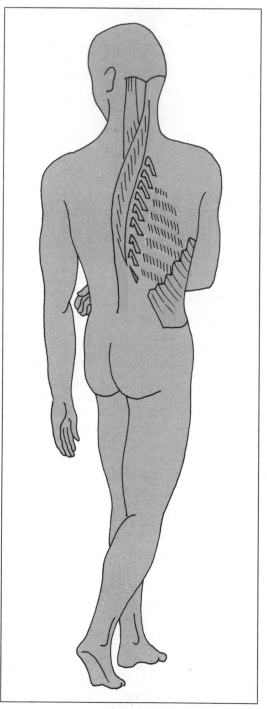

Fig 4–4.—The muscles participating in rotating the trunk. Dorsal view after removing the superficial muscle layer and the scapula (after *Benninghoff*).

- The *deltoid m.* (9.9 mkp) (This muscle does most of the work.)
- The short head of the *biceps brachii m.* (1.7 mkp)
- The *supraspinatus m.* (1.4 mkp)
- The *pectoralis major m.* (0.8 mkp)
- The *infraspinatus m.* (0.8 mkp)
- The *coracobrachialis m.* (0.7 mkp)
- The *subscapularis m.* (0.6 mkp)

The total combined power developed by the muscles in raising the arm anteriorly is approximtely 17 mkp, even though only parts of each muscle participate. It is evident that while the *deltoid m.* does most of the work, the total work generated by individual smaller muscles is still considerable.

Elevation

The movement of the arm from the anterior horizontal position to the vertical is known as "elevation."

Note: This movement is made possible only by the rotation of the scapula!

Participating Muscles (Fig 4–5):

- The *deltoid m.*

Thc higher the elevation, the more and more components of this muscle participate (see p. 41 and Fig 2–7)

- The *serratus anterior m.*

This muscle pulls the inferior angle of the scapula anteriorly and, thus, makes the movement of the arm from the horizontal to the vertical possible.

- The *trapezius m.*

The upper portion of this muscle contributes to the rotation of the shoulder by raising the shoulder. The lower portion of the *trapezius m.* acts by pulling the superior angle of the scapula downward.

The failure of any one of the above muscles makes elevation impossible.

The muscles are particularly prominent in weight lifters, since in the "jerk" they must be exerted to their maximal power.

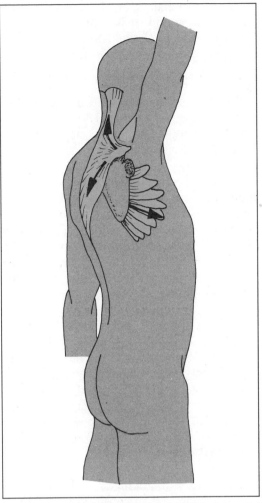

Fig 4–5.—The muscles participating in elevating the arm and rotating the scapula.

The Fixation of the Arms in the Elevated Position

In anterior flexion and subsequent elevation, the arm can be raised to the vertical position. Bony obstruction in the shoulder prevents any further movement of the arm in this direction. To fix the arm in this position (e.g., in handstands), both the bony obstruction and, even more so, the active contribution of the muscles, is necessary. Both those muscles that elevate the arm and those that lower it (see below) must participate in this maneuver. It is thus the isometric contraction of both the agonists and antagonists that accomplishes the fixation of the arm in the elevated position.

Lowering the Arm From the Vertical

When standing, this is accomplished by gravity alone. The gradual lowering of the arm is accomplished by the stepwise relaxation of the tone of the muscles that produce the elevation.

In an active counter move (e.g., front support mount on the horizontal bar) or in a forceful acceleration of the move (e.g., in all throwing or hitting maneuvers) the arm-lowering muscles dominate totally. The tone of the above-mentioned antagonists is reduced to a minimum, in order not to hinder the agonists in developing their full power.

Participating Muscles

The muscles acting on the anterior aspect of the trunk or on the shoulder joint:
- The *pectoralis major m.*

This muscle is *the* throwing muscle par excellence. Its maximum power is developed from the vertical position down to the horizontal sphere.
- The long head of the *biceps brachii m.*

This muscle participates along the entire arc, from the vertical position above to the vertical position below.

The muscles acting on the posterior aspect of the trunk and from the scapula on the shoulder joint.
- The *latissimus dorsi m.*

This muscle and the *pectoralis major m.* are *the* muscles of throwing and pulling!
- The *teres major and minor muscles*

The *teres major m.* is, above all, a powerful lowering muscle of the arm, which participates in the entire arc of acceleration from top to bottom (particularly important in swimming)
- The *subscapularis m.*

In addition, all muscles that rotate the scapula posteriorly participate indirectly in the arm-lowering maneuver:
- The *rhomboideus major and minor muscles*
- The *trapezius m.*

The descending portion of this muscle, which inserts on the superior angle of the scapula, acts as an antagonist to the ascending portion

Retroversion of the Arm (From the Anatomic Position)

The retroversion of the arm is a continuation of the movement described above. It is limited and can develop relatively little power. It plays a role in cross-country skiing (arm action) and in free-style swimming (at the end of the power stroke and at the beginning of the recovery phase).

Participating Muscles

- The *deltoid m.* (0.9 mkp)
- The *subscapularis m.* (0.9 mkp)
- The *teres major m.* (0.8 mkp)
- The *latissimus dorsi m.* (0.3 mkp)
- The *triceps brachii m.* (0.1 mkp)

The retroversion of the arm is accomplished primarily by the first three muscles that arise from the region of the shoulder. To improve their pulling power, the lower angle of the scapula is rotated backward by the *rhomboideus major and minor muscles* and by the *trapezius m.*

Abduction of the Arm (From the Anatomic Position)

A strong abduction is important particularly in weight lifting, in the "raising" phase.

Participating Muscles

- The *deltoid m.* (10.4 mkp)

The acromial portion of this muscle is the primary agonist in this maneuver
- The *infraspinatus m.* (2.7 mkp)
- The *supraspinatus m.* (2.4 mkp)
- The long head of the *biceps brachii m.* (1.1 mkp)

Volleyball–National League: A
smash hit by Schafer (13) is
blocked by Buhner (10) and
Rath (3).

Retroversion of the Arm From Abduction

Retroversion of the arm from the abducted position, (gymnastics and wind-up in discus throwing) is usually part of an external rotatory movement. During this maneuver, the *infraspinatus m.*, the *teres minor m.*, and the *deltoid m.* move the arm toward the scapula, while the *trapezius* and *rhomboideus major and minor muscles* move the scapula toward the spinal column.

Anteversion From the Lateral to the Anterior Position

The maneuver can be performed with remarkable power by the *pectoralis major m.*, as in discus throwing. It is assisted by the *deltoid m.*, the short head of the *biceps brachii m.*, and the *coracobrachialis m.*

Adduction of the Arm From the Abducted Position

Moving the arm from abduction toward the trunk and the fixation of the arm in the abducted position, while forcefully trying to lower it, is by far the most powerful movement of the shoulder joint (total power = approximately 40 mkp). This maneuver is seen in the "iron cross hang" on the still rings.

In performing the adduction maneuver, the scapula and the shoulder girdle must be fixed to the thorax by the *trapezius m.*, the *rhomboideus major and minor muscles*, the *serratus anterior m.*, and the *pectoralis major m.* This fixation is essential to provide the muscles, which originate from this region, with a solid base.

Participating Muscles

Muscles acting from the anterior trunk wall on the shoulder girdle and from the shoulder girdle on the shoulder joint:
- The *pectoralis major m.* (11.8 mkp)
- The *triceps brachii m.* (8.5 mkp)

Besides these two primary agonists we find:

- The *deltoid m.* (3.4 mkp)
- The short head of the *biceps brachii m.* (2.1 mkp)
- The *coracobrachialis m.* (2.0 mkp)

The muscles acting on the shoulder joint from the posterior aspect of the trunk and from the scapula:
- The *teres major m.* (7.3 mkp)
- The *latissimus dorsi m.* (5.5 mkp)
- The *subscapularis m.* (1.0 mkp)

Internal Rotation of the Arm From the Neutral Position

This maneuver is important in the martial arts (judo and wrestling), in fencing and swimming (in the transition from the pull to the push phase in free style, back stroke, and butterfly breast stroke).

Participating Muscles

- The *subscapularis m.* (3.3 mkp) (This muscle is the most powerful internal rotator of the arm!)
- The *pectoralis major m.* (1.0 mkp)
- The long head of the *biceps brachii m.* (1.0 mkp)
- The *teres major m.* (0.8 mkp)
- The anterior portion of the *deltoid m.* (0.3 mkp)
- The *latissimus dorsi m.* (0.3 mkp)

External Rotation of the Arm From the Neutral Position

The external rotation is important in the martial arts, in fencing, and in swimming (breast stroke).

Participating Muscles

- The *infraspinatus m.* (2.5 mkp) (This muscle is the primary agonist in this maneuver.)
- The *deltoid m.* (posterior aspect) (0.4 mkp)
- The *teres minor m.* (0.3 mkp)

The total power output in external rotation is only about half of that developed in internal rotation. For this reason, the arm is slightly internally rotated in the neutral, anatomic position, since the tone of the internal rotators is dominant.

Flexion of the Arm at the Elbow Joints

The flexors play an important role in all movements involving carrying and in the pulling or climbing maneuvers.

Participating Muscles

- The *biceps brachii m.* (4.8 mkp)
- The *brachialis m.* (3.8 mkp)
- The *brachioradialis m.* (1.9 mkp)
- The *pronator teres m.* (1.2 mkp)
- The *extensor carpi radialis longus m.* (1.2 mkp)

All muscles that act on the wrist joint and also have an incidental flexor activity on the elbow joint must be included in the above (0.9 mkp).

The flexors of the arm are differentially effective in pronation and supination. The *biceps brachii m.* and *the brachialis m.*, which is located under the biceps, function best with the forearm in supination, while the *coracobrachialis m.* can exert its maximum power with the forearm in the neutral or pronated position.

Extension of the Arm at the Elbow Joint

Participating Muscles

- The *triceps brachii m.* (8.5 mkp)
- The *anconeus m.* (0.8 mkp)

The main extensors of the elbow joint are the two short heads of the *triceps m.* (6.1 mkp), which act on the elbow joint alone. The long head of the *triceps m.*, which acts also on the shoulder, is less important.

Since the total motive power of the flexors is greater than that of the extensors, the relaxed arm usually assumes a slightly flexed position.

The *triceps m.* is extraordinarily important in all athletic activities in which the extension of the arm and the fixation of the arm in extension is required (shot put, javelin throw, boxing and gymnastics, etc.)

Rotatory Movements of the Elbow Joint

Supinators

- The *biceps brachii m.* (1.1 mkp)
- The *supinator m.* (0.3 mkp)

Support is given by several other muscles (0.3 mkp). The total power output of the supinators is greatest when the elbow is flexed at right angles!

Pronators

- The *pronator teres m.* (0.7 mkp)
- The *pronator quadratus m.* (0.2 mkp)

To this must be added a number of other muscles, which have pronation as a side effect (0.6 mkp).

The total power of the pronators approximately equals the power of the supinators, except that the pronators function best with the arm extended and the supinators when the elbow is flexed at right angles (see p. 91).

In athletics, the supinators and pronators are important, albeit to different degrees, in the martial arts (judo, wrestling) in fencing, and in swimming.

Flexion of the Wrist (From Extension)

Participating Muscles

- The *flexor digitorum superficialis m.* (4.8 mkp)
- The *flexor digitorum profundus m.* (4.5 mkp)

The two long flexor muscles of the fingers are also the most powerful flexors of the wrist joint.

- The *flexor carpi ulnaris m.* (2.0 mkp)
- The *flexor pollicis longus m.* (1.2 mkp)
- The *flexor carpi radialis m.* (0.8 mkp)

The total potential power output is approximately 13 mkp.

The reason for this remarkable strength becomes obvious when we examine the role of these muscles in athletics. In gymnastics, throwing and shot put, and in the martial arts, great demands are placed on these muscles. The flexors of the fingers have, above all others, important dynamic functions (e.g., pushing off in vaulting and shot put) and also static functions, such as rigid positions in gymnastics (single or double handstand), rowing, weight lifting, tennis, the martial arts, etc.

Extension of the Wrist Joint (From the Flexed Position)

Participating Muscles

- The *extensor digitorum communis m.* (1.7 mkp)
- The *extensor carpi radialis longus m.* (1.1 mkp) and brevis (0.9 mkp)
- The *extensor indicis m.* (0.5 mkp)

The power of the extensors of the wrist, and equally of the extensors of the fingers, is significantly less than the power of the flexors. This is due to the fact that the hand is primarily a grasping and holding tool and that the range of functions that depends on the extensors is significantly less.

In fencers, wrestlers, tennis players and, particularly, in weight lifters (lifting maneuvers), the extensors will be well developed, since in these sports the fixation of the wrist joint requires an increased force of contraction.

Abduction of the Wrist Joint

Ulnar Abduction

- The *extensor carpi ulnaris m.* (1.1 mkp)
- The *flexor carpi ulnaris m.* (0.7 mkp)

Radial Abduction

- The *extensor carpi radialis longus m.* (1.1 mkp) and *brevis* (0.3 mkp)
- The *abductor pollicis longus m.* (0.4 mkp)
  Several other muscles also participate (0.3 mkp)

Radial abduction plays a role in the final step in discus throwing; ulnar abduction in the breast stroke. In the martial arts, the ab-

ductors are necessary in fixing the wrist joint when both flexors and extensors are contracting (e.g., hitting with the edge of the hand, all blows in boxing).

Simple Motions of the Lower Extremity

Hip Flexion (Anterior Flexion of the Thigh From the Anatomic Position)

Participating Muscles (Fig 4–6):

- The *rectus femoris m.* (16.4 mkp)
- The *iliopsoas m.* (10.0 mkp)
- The *tensor fasciae latae m.* (7.5 mkp)
- The *sartorius m.* (4.3 mkp)
- The anterior portion of the *gluteus minimus m.* (3.5 mkp)
- The *pectineus m.* (2.7 mkp)

From the retroverted position, the abductors also participate in this motion.

The total energy output is approximately 45 mkp.

The extent of the anterior flexion depends not only on the contractile force of the hip flexors, but also, and to a significant degree, on the position of the knee and on the associated stretch of the *ischiocrural muscles*. When the leg is anteverted with the knee extended, these muscles evince a strong resistance to stretch. This allows significantly less flexion than when the knee is flexed and these muscles are much less stretched.

The high incidence of injury in the dorsal aspect of the thigh in soccer players is due to the fact that kicking a ball toward the net entails an extreme stretching of these muscles during the explosive bending of the hip and the simultaneous extension of the knee. When the warm-up period is too short, or the stretching exercises before the game are inadequate, or when the player becomes excessively tired, muscle tears can occur quite readily.

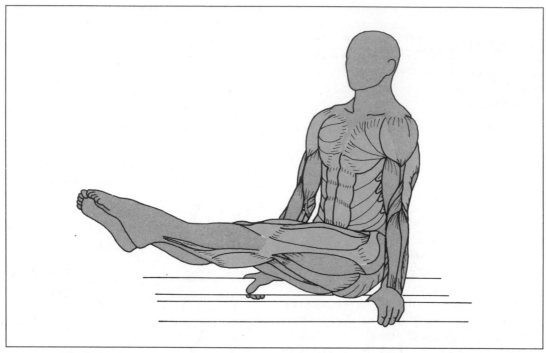

Fig 4–6.—The muscles participating in flexing the hip in the ''L'' support maneuver.

Extension of the Hip (From Flexion Into the Anatomic Position)

Participating Muscles

- The *gluteus maximus m.* (53.2 mkp)
- The *adductor magnus m.* (22.2 mkp)

It is evident that this adductor, in addition to its true function, also plays an extraordinarily important role when the support column (leg), has been moved from the center of gravity of the body, and must be returned again to the line of support.

- The *semimembranosus m.* (17.0 mkp)
- The *semitendinosus m.* (7.0 mkp)
- The posterior portion of the *gluteus medius m.* (6.0 mkp)
- The long head of the *biceps femoris m.* (4.4 mkp)
- The *quadratus femoris m.* (3.4 mkp) and a further group of muscles in the region of the hip. The total energy output is approximately 120 mkp!

It can be seen that the primary extensor of the hip, the *gluteus maximus m.*, receives appreciable support in hip extension, not only from the *adductor magnus m.*, but also from the *ischiocrural* group of muscles (*semitendinosus m., semimembranosus m., and biceps femoris m.*).

The extraordinarily great strength of the hip extensors is due to their importance in maintaining the erect posture and in forward motion. In athletics, the extensors are of critical importance in all accelerated maneuvers from the previously flexed hip position (e.g., rising from the squatting position in weight lifting) and also in reestablishing or maintaining the balance (e.g., when landing after a jump, such as ski jumping).

Overextension of the Hip (Retroversion of the Thigh)

- The *gluteus maximus m.* (10.4 mkp)
- The *gluteus medius m.* (5.7 mkp) and other muscles (see above)

Total energy output, 22 mkp

Because of the already significant shortening of the muscles, and consequently, because of the unfavorable conditions for further contraction, the extensors of the hip can perform this motion only quite poorly. In fact, they are almost entirely outside their normal line of traction (e.g., the powerful *adductor magnus m.*).

Furthermore, the overextension of the hip is limited by a very strong system of ligaments, particularly the iliofemoral ligament. Only when the trunk is flexed forward and the "ligamentous helix" is loosened can an increased retroversion of the thigh take place.

Abduction of the Thigh (From the Anatomic Position)

Participating Muscles

- The *gluteus medius m.* (12.4 mkp)
- The *rectus femoris m.* (9.8 mkp)

Since during every effort to spread the legs, this biarticular muscle comes into play in the region of the axis of abduction, it not only forcefully flexes the hip, but also functions as an abductor.

- The *gluteus maximus m.* (iliotibial tract insertion) (9.6 mkp)
- The *tensor fasciae latae m.* (8.6 mkp)
- The *gluteus minimus m.* (7.1 mkp)
- The *sartorius m.* (1.9 mkp)
- The *piriformis m.* (1.6 mkp)

The total energy output of the abductors: approximately 51 mkp.

The great strength of the abductors is of extraordinary importance from a static point of view. In walking, the abductors tilt the pelvis toward the standing leg and thus allow the free forward motion of the swinging leg. When the abductors are unable to fixate the pelvis toward the side of the standing leg, the unsupported side of the pelvis descends during the standing phase of the injured leg. The necessary balancing movements of the trunk, which must lean toward the standing leg-side to allow the forward motion of the swinging leg, will lead to the development of a waddling gait (Fig 4–7).

The thigh can be abducted more strongly

when the hip is flexed. Just as in retroversion, the loosening of the "ligamentous helix" is of importance in this maneuver also.

Adduction of the Thigh (From the Abducted Position to the Anatomic One)

Participating Muscles

- The *adductor magnus m.* (28.0 mkp)
- The femoral insertion of the *gluteus maximus m.* (12.5 mkp)
- The *adductor longus m.* (12.2 mkp)
- The *adductor brevis m.* (9.0 mkp)

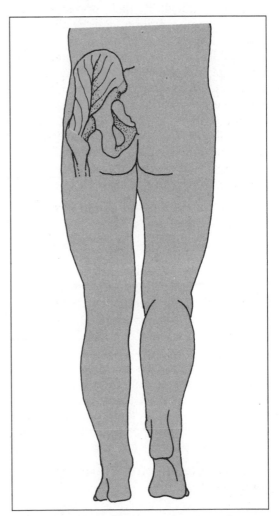

Fig 4–7.—The role of the gluteus maximus m. and of the adductor muscles in walking on the side of the standing leg.

- The *semimembranosus m.* (8.4 mkp)
- The *ilipsoas m.* (5.8 mkp)
- The long head of the *biceps femoris m.* (5.5 mkp)
- The *semitendinosus m.* (3.9 mkp)
- The *pectineus m.* (3.7 mkp)
- The *obturator externus m.* (3.7 mkp)
- The *gracilis m.* (2.9 mkp)
- The *quadratus femoris m.* (2.2 mkp)

The total energy output of the adductors is approximately 100 mkp, of which half is contributed by adductor muscles.

The extraordinary strength of the adductors is explained by their primarily static function. In cooperation with the abductors, they are largely responsible for maintaining the equilibrium of the pelvic posture in the frontal plane. When this muscular balance is disrupted, the ability to maintain the erect posture and forward motion is markedly interfered with.

In athletics, the adductors are particularly important in riding, in skiing, and in the major team sports where rapid directional movement changes are required.

Internal Rotation of the Thigh (From the Anatomic Position)

Participating Muscles

- The *adductor magnus m.* (no mkp)
- The *adductor longus m.* (no mkp)
- The *tensor fasciae latae m.* (0.9 mkp)
- The *gluteus maximus m.* (0.8 mkp)
- The *rectus femoris m.* (0.5 mkp) and several other muscles

Internal rotation of the thigh is the least powerful movement of the hip joint. Flexing the hip improves the effective lines of traction and improves the power of this maneuver, which is important in alpine skiing (snow plowing).

External Rotation of the Thigh (From the Anatomic Position)

Participating Muscles

- The *gluteus maximus m.* (9.0 mkp)
- The *gluteus medius m.* (4.2 mkp)

- The *triceps coxae* (the *obturator m.* and the *two gemelli m.*) (2.5 mkp)
- The dorsal femoral insertion of the *adductor magnus m.* (2.2 mkp)
- The *rectus femoris m.* (1.4 mkp)
- The total energy output is approximately 30 mkp

External rotation can be performed with approximately the same force than the other hip movements. Contrary to the internal rotation and abduction, however, flexion of the hip does not increase the strength of external rotation.

When the foot is relaxed (swing leg— in walking), the foot is rotated slightly to the outside, since the tone of the external rotators is greater than the tone of the internal rotators.

In athletics, external rotation is important in the martial arts, in figure skating, in soccer (kick with the inside of the foot), and in discus throwing (beginning of the rotation).

Extension of the Knee Joint

Participating Muscles

- The *quadriceps femoris m.* (142 mkp!)

The biarticular *rectus femoris m.* contributes 23.4 mkp to the total energy output

- The *tensor fasciae latae m.* (0.8 mkp)

The *quadriceps femoris m.* can determine the upper limits of performance in all athletic maneuvers where extension of knee joint is essential. These include all jumping and running, weight lifting (rising from the squatting position), etc.

When the *quadriceps femoris m.* is stretched by the extension of the hip joint, it can extend the knee joint more forcefully. Flexion of the hips, on the other hand, reduces the prestretching of the *rectus femoris* and thus reduces the strength of the muscle. For this reason, sit-ups are more difficult when the knees are flexed since, in this position, the *rectus femoris* can assist, albeit poorly, in flexing the hip.

Flexion of the Knee Joint

Participating Muscles

- The *semimembranosus m.* (16.8 mkp)
- The *semitendinosus m.* (13.2 mkp)
- The *biceps femoris m.* (10.3 mkp)
- The *gracilis m.* (3.1 mkp)
- The *sartorius m.* (2.3 mkp)
- The total energy output is approximately 46 mkp

Since the *ischiocrural* muscles are biarticular, and since they frequently have opposing functions in the same maneuver, they are prone to injury.

Since the ischiocrural muscles can extend the hip and flex the knee (Fig 4–8) they have to contract to extend the hip in running and extending the support leg, and to stretch to extend the knee at the same time. This simultaneous contraction and stretching can easily result in muscle tears and other injuries, particularly in the absence of adequate warm-up and stretch exercises. The swinging leg is less frequently involved in injury, since here the flexion of the knee during flexion of the hip tends to be passive, rather than active, even though the mechanism is essentially the same.

Another typical mechanism that produces injury is to bend the hip forcibly forward, while the knee is extended (e.g., in sliding tackles in football). In this maneuver, the *ischiocrural* muscles are overextended.

Internal Rotation of the Leg

Participating Muscles

- The *semimembranosus m.* (3.4 mkp)
- The *semitendinosus m.* (0.8 mkp)
- The *popliteus m.* (0.8 mkp)
- The *sartorius m.* (0.6 mkp)
- The *gracilis m.* (0.4 mkp)
- The total energy output is approximately 6 mkp

Internal rotation of the leg is possible only when the knee is flexed and is limited even then (10°). This fact is important in

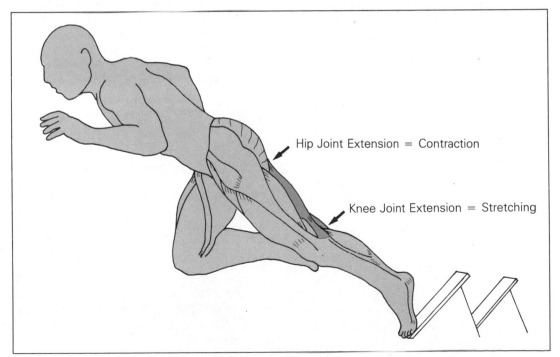

Hip Joint Extension = Contraction

Knee Joint Extension = Stretching

Fig 4–8.—Simultaneous relaxation and contraction of the ischiocrural muscles in running (support leg).

Alpine skiing to maintain the "knee action."

External Rotation of the Leg

Participating Muscles

- The *biceps femoris m.* (4.9 mkp)
- The *tensor fasciae latae m.* (0.6 mkp)

External rotation is also possible only when the knee is flexed (to 40°).

Plantar Flexion in the Upper Ankle Joint

Participating Muscles

- The *gastrocnemius m.* (9.0 mkp)
- The *soleus m.* (7.4 mkp)
- The *flexor hallucis longus m.* (0.9 mkp)
- The *flexor digitorum longus m.* (0.4 mkp)
- The *tibialis posterior m.* (0.4 mkp)
- The *peroneus longus m.* (0.4 mkp)
- The *peroneus brevis m.* (0.3 mkp)

The *triceps surae m.*, consisting of the *gastrocnemius m.* and *soleus m.*, produces 9/10 of the total energy output in plantar flexion and is, thus, for all practical purposes *the* muscle of forward motion. It gives acceleration to 97% of the body mass and, for this reason, must be strongly developed.

The other five muscles are primarily positional muscles, since their shortness, unfavorable leverage, and small diameter can develop only limited force. Nevertheless, they are essential, since they position the foot correctly for placement on the ground and thus enable the *triceps surae m.* to develop full power.

How weak these muscles are becomes obvious when there is a reversal between the fixed point and the moving point. If, for example, the foot is caught in a hole in the ground and the forward momentum of the body acts on these muscles, the muscles are too weak to act as an efficient brake and as a result an excessive demand is placed on the passive system of locomotion, the tendons and ligaments. This can easily lead to

a stretching and even to a tearing of the ligaments.

In walking or standing, the plantar flexors of the foot position the sole of the foot flatly against the ground and thus provide the support-column function of the lower extremity with the largest possible base.

Dorsal Extension of the Upper Ankle Joint

Participating Muscles

- The *tibialis anterior m.* (2.5 mkp)
- The *extensor digitorum longus m.* (0.8 mkp)
- The *peroneus tertius m.* (0.5 mkp)
- The *extensor hallucis longus m.* (0.4 mkp)
- The total energy output is approximately 4 mkp

The main function of the dorsal extensors is in walking. They raise the foot to enable the swinging leg to move forward without hindrance. On the standing leg side, they pull the calf toward the fixed foot, a reversal of the fixed and mobile points. They are important in Alpine skiing and in cross-country skiing. It is not surprising that after long marches these muscles become painful ("charley horse").

Pronation of the Foot in the Lower Ankle Joint

Participating Muscles

- The *peroneus longus m.* (1.1 mkp)
- The *peroneus brevis m.* (0.9 mkp)
- The *extensor digitorum longus m.* (0.5 mkp)
- The *peroneus tertius m.* (0.4 mkp)

More than half of the energy output of the pronators is provided by the two lateral calf muscles, the *peroneus longus and brevis m.* They play a role in positioning the foot, particularly on the lateral side (they adapt the foot to uneven ground surfaces and changing ground contours). They are important in alpine skiing, particularly in "setting the edges."

Supination of the Foot in the Lower Ankle Joint

Participating Muscles

- The *gastrocnemius m.* (2.5 mkp)
- The *soleus m.* (2.3 mkp)
- The *tibialis posterior m.* (1.5 mkp)
- The *flexor hallucis longus m.* (0.7 mkp)
- The *flexor digitorum longus m.* (0.6 mkp)
- The *tibialis anterior m.* (0.3 mkp)

In conjunction with the pronators, these muscles serve to position the foot optimally. In athletics, they are important in those maneuvers that require a particularly precise positioning of the foot, e.g., in gymnastics (on the balance beam) and in figure skating.

Besides the already described movements, the upper and lower ankle joint can also perform abduction and adduction. These are largely connected with supination and pronation and will not be further discussed.

5 The Analysis of Complex Athletic Maneuvers

Introduction

A systematic discussion of almost all of the Olympic athletic events should provide the reader with instant information concerning his/her particular athletic interest. If additional information is required, a review of the previously discussed individual muscles or of the analysis of simple trunk and extremity movements should be helpful.

The illustrations are limited to those muscles that are relevant to the specific movement. The particular athletic maneuver is depicted at the peak of the muscular contractions (muscles shown in red).

Since the so-called ''track and field'' athletic events arc fundamental to many athletic and sporting events, and since the elements of motion that are characteristic of ''track and field'' athletics are used in most of the other events directly, or only with slight modifications, these events will be discussed in particular detail.

Track and Field

The same muscle groups are used in walking, running, and jumping. The differences lie only in the increasing mobilization of the different muscle groups.

Walking

In walking, we must distinguish an anterior and posterior stance phase (standing leg) and an anterior and posterior swing phase (swing leg). The complete motion cycle of walking and the muscular activity on which it is based shall be examined in the context of a double step (Fig 5–1).

The Muscles Active in the Swing Phase

The Posterior Swing Phase

The swing phase of the swing leg begins when the rear leg pushes off, using primarily the *tricep surae m*. This pushing off as-sists in establishing the lever action of the anterior, support leg. During the posterior swing phase, the contraction of the *ischiocrural muscles,* present during the preceding anterior and posterior support phase, continues. This leads to a flexion of the knee as soon as the leg no longer has to bear weight (see Fig 5–1, b). While the thigh assumes the vertical position during the posterior swing phase, by gravity, the leg is slightly lifted by the *ischiocrural muscles,* in a forward swinging motion. This forward swing of the leg is accompanied by a simultaneous dorsal extension of the foot in the superior ankle joint (mostly by the *tibialis anterior m.*).

The Anterior Swing Phase

As the hip is increasingly flexed (raising of the thigh), by the contraction of the *rec-

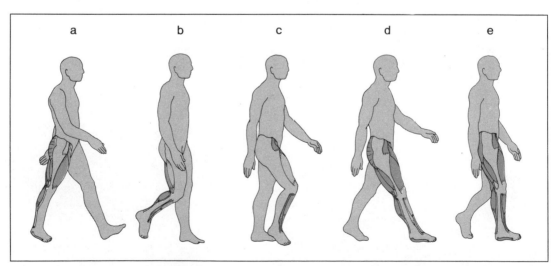

Fig 5–1.—(**a-e**) The muscles participating in the posterior and anterior swing and stance phases in walking.

tus femoris m., iliopsoas m., tensor fasciae latae m., and others (see Fig 5–1, c), the ischiocrural muscles are increasingly stretched. This results in an increasing flexion of the knee, corresponding to an increasing elevation of the thigh. This passive maneuver assures the correct placement of the foot onto the ground. At the end of the forward swing phase, the leg extension takes place. In slow walking and short steps, this extension of the leg takes place by gravity alone. In fast walking and increased stride, this extension requires the active participation of the quadriceps femoris m.

At the end of the forward swing phase, the heel of the swing leg makes contact with the ground, after this phase has been stopped by the ischiocrural muscles. At this point, the anterior stance phase of the supporting or standing leg begins.

The Anterior Stance Phase

Placement of the heel triggers the contraction of the gluteus maximus m. In its role as extender of the hip joint, it is assisted by the adductor and ischiocrural muscles (see Fig 5–1, d).

The leg is flexed toward the foot, primarily by the tibialis anterior m. The triceps surae m. positions the sole of the foot on the ground and achieves its maximal contraction at the moment when the toes touch the ground. The positional muscles of the foot assist in this maneuver.

The stabilization of the leg as support column is the responsibility of the quadriceps femoris m., which has its maximum contraction when the ball of the foot touches the ground, and of the triceps surae m., which, through its iliotibial tract, accomplished the extension and stabilization of the knee joint.

When the support leg passes beyond the vertical position, the posterior stance phase begins (see Fig 5–1, e).

The Posterior Stance Phase

The full extension of the hip, to the point of "pushing-off," is accomplished by the already mentioned gluteus maximus m. and the ischiocrural muscles. At the moment when the foot "pushes off," all the hip extenders, knee extenders (quadriceps femoris m.) and plantar flexors (triceps surae m. and other flexors) achieve their second and maximal contraction (see Fig 5–1, a).

In summary, walking is accomplished by a series of muscular contractions which, acting on different joints, are tuned to cooperate delicately and transform the components of walking into smoothly orchestrated gliding forward motion.

The Muscles that Determine Performance

The length and frequency of the steps determines the speed of walking. In the swing phase these are determined by the flexors of the hip joint (the rectus femoris m., the iliopsoas m., the tensor fasciae latae m.). In the stance and pushing-off phase, the extenders of the hip (gluteus maximus m.), and the extensors of the knee (quadriceps femoris m.) and plantar flexors, (triceps surae m.) are dominant.

The Start and Sprint

Even though at the start, from the "get set" position, the arms give some support, and the contralateral swinging of the arms helps during the sprint, nevertheless, the brunt of the forward motion is carried by the musculature of the lower extremity. For this reason, the subsequent discussion highlights this area.

The "get set" position serves to prestretch the running muscles and to place them at an optimal angle of effect. The most important item is to stretch the triceps surae m. sufficiently, by pushing the heel forcibly against the starting block. This will increase the pushing-off force and also the initial acceleration, by virtue of the fact that the whole foot is braced against the entire starting block (Fig 5–2).

The firing of the starting gun triggers the extension of the anterior "starting" leg by a simultaneous extension of the hip and the knee as well as the terminal palmar flexion of the foot by the *triceps surae m.* The explosive flexion of the hip on the swing leg side is accomplished primarily by the *rectus femoris m.,* the *iliopsoas m.,* and the *tensor fasciae latae m.*

It is easy to see that starting and sprinting uses the same muscles as walking. The distinguishing factor in sprinting is the significantly increased total performance, and the dramatically increased work of acceleration and speed.

There are a number of complex exercises to strengthen the extensors of the hip and knee joints and of the plantar flexors e.g., squat jumps, step-up exercises with terminal push-offs, etc. There are also selective exercises to strengthen individual muscle groups, e.g., raising and lowering the heel while standing on the edge of a step (Fig 5–3, a) *(triceps surae m.),* flexion of the leg against resistance (the *ischiocrural m.*) (Fig 5–3, b) and straightening of the legs in the leg exercise machine (Fig 5–3, c).

Broad Jumping

Jumping-off uses the same performance muscle groups as sprinting.

In preparation for the landing, a special conditioning of the hip flexors and of the abdominal muscles becomes necessary. The flexors of the hip make the required raising of the legs possible, while the abdominal muscles make their contribution by fixing the pelvis and by rotating it backward.

Since flexion of the hip is affected by the stretch resistance of the biarticular *ischiocrural* muscles, it becomes absolutely essential for optimal performance to train these muscles for improved stretch potential (Fig 5–4).

The Triple Jump

The musculature that determines performance in the "hop," "step," and "jump," triple jump corresponds exactly to muscles used in the broad jump. The twofold, intermediate landing, however, requires a strengthening of the muscles that stabilize

Fig 5–2.—Stretching the support leg and explosive forward motion of the swinging leg (**b**) from the "get ready" position (**a**).

At the World Cup races,
Williams wins just ahead of
Ray in the 100 m-sprint.

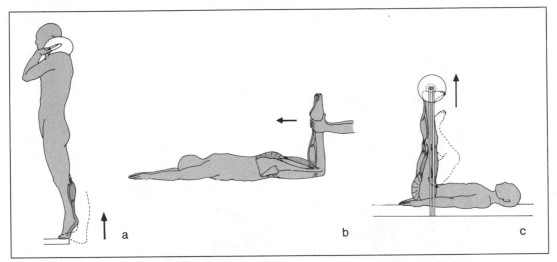

Fig 5–3.—Selective exercises for the plantar flexors: **(a),** the knee flexors, **(b)** and **(c),** the knee extensors.

the hip and the thigh to prevent a forward tilt of the pelvis.

At the moment of landing (Fig 5–5, a and b), the jumping muscles act in an antidynamic fashion, i.e., they neutralize and distribute the effects of the preceding jump and relax. This is followed by a positive dynamic action (Fig 5–5, c), i.e., they contract. This combination requires special training exercises (repeated multiple jumps, jumping from a platform, etc.). At the same time the muscles of the hip must also stabilize the legs. It is important to provide special training to strengthen the abductors (see p. 105) and the adductors (see p. 103) in excess of normal conditioning. This can be accomplished by exercises such as the lateral lift of the trunk when lying on the side and the legs are fixed, or raising the legs when the trunk is fixed (abductor training).

High Jump

When jumping off, the same muscles are used as in sprinting, broad jump, and triple jump, although there must be an increase in the acceleration force, since in this event the entire body must be raised and acceler-

ated from the horizontal into the vertical position.

To extend the swing leg in the "straddle" and to arch the back in the "flop" requires great contractile force of the agonists (flexors of the hip and extensor of the back) and great elasticity of antagonists (particularly the *ischiocrural* and abdominal muscles) (Fig 5–6).

Pole Vault

The pole vault is a very complicated maneuver. The different phases of this vault make very high demands of the muscles of the legs, trunk, and arms.

The Muscles that Determine Performance

The Jumping-off Phase

The activation of the spring leg and swing-leg (see broad jump and sprint).
- Activation of the arms
- The lower or pushing arm: *Triceps brachii m.*
- The upper or pulling arm: primarily the *pectoralis major m.* (see p. 138) and also the flexors of the arm *(biceps brachii, brachialis, brachioradialis m.)*

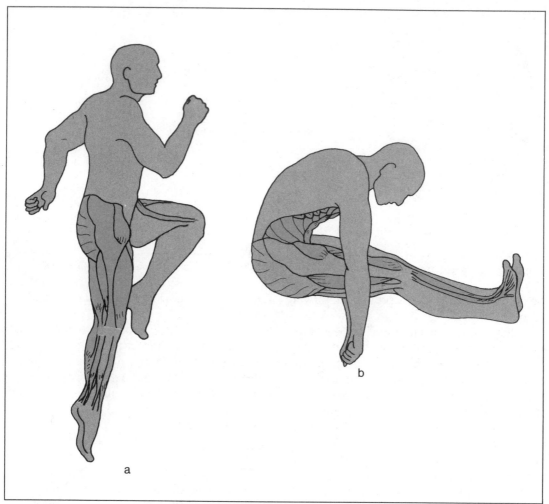

Fig 5–4.—Extension of the jumping leg and deployment of the swinging leg in the broad jump: (**a**), the muscles participating in preparation for landing (**b**).

Swing and Roll

The work of the pushing and pulling arms continues. Since the trunk must be approximated to the arms, the adductors of the arms are activated (reversal of the fixed and mobile points). The pelvis must also be brought closer to the arms and this is accomplished by a contraction of the abdominal muscles and the hip and thigh flexors (Fig 5–7).

As a particularly useful strengthening exercise for this phase of the vault, we rec-ommend repeated hang-to-support exercises on the horizontal bar and repeated leg lift exercises from the hang position to the horizontal bar. This will train the flexors of the hip, the abdominal muscles, and the arm-lowering muscles all at the same time.

Extension and Rotation

Extension of the pole is accompanied by a simultaneous extension of the hip (*gluteus maximus m.* assisted by the *ischiocrural m.*), the trunk *(erector spinae m.)* and the

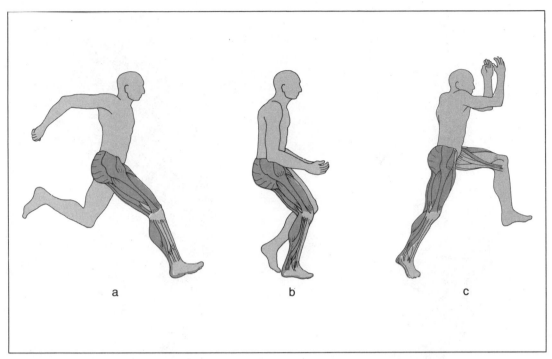

Fig 5–5.—Triple jump: The jumping and equalizing muscles and the stabilizer muscles of the hip that are quite particularly involved in the landing after the ''hop'' (**a** and **b**) and the take off for the ''step'' (**c**).

knee *(quadriceps femoris m.)*. The flexed lower arm is pulled toward the hips *(biceps brachii, brachialis, brachioradialis m.)*, which approximates the trunk to the pole.

When the hip reaches the level of the upper hand, rotation along the longitudinal axis begins (rotatory muscles of the trunk, p. 135). While the upper arm begins to flex, the lower arm extends.

Letting Go and Crossing the Bar

When the shoulder of the upper arm rises higher than the arm itself, the lower arm begins to push off from the pole, and the upper arm quickly follows suit. In both instances, the *triceps brachii* is the active muscle.

Special Strength Training

For the extension and rotation of the trunk: kip exercise into a handstand (with or without a half turn), backward roll with mo-

mentary handstand. For the flexors of the arm: climbing a rope, chin-ups. For the extensors of the arm: walking on the hands, jumping while in a handstand position, and push-ups.

Javelin Throwing

The Muscles that Determine Performance (Fig 5–8)

Leg action: Javelin throwing requires great strength of the hip extensors, the knee extensors, and the plantar flexors. This is required both for the bracing action and the extension of the leg.

Trunk action: The rotators of the trunk (see p. 135) and the abdominal muscles must be particularly strong to sustain the throwing motion from the wind-up (see Fig 5–8, a), through the body arch (see Fig 5–8, b), to the forward whipping motion of the trunk (see Fig 5–8, c). In the last phase, the flex-

Fig 5–6.—The muscles participating in the taking-off phase of the "straddle."

ors *(rectus femoris, iliopsoas, tensor fasciae latae muscles)* of the hip play a major role.

Arm action: The forward, whipping movement is accomplished primarily by the arm-lowering muscles *(pectoralis major* and *latissimus dorsi m.)* and the extensors of the arm *(triceps brachii m.).*

Conjoint exercises of abdominal muscles, hip flexors, and throwing arm muscles: throwing a beach ball toward the ceiling in a supine position (the hips and back lie flat on the transversely placed training table with the shoulder and arm hanging over the end and a colleague holding the legs down).

Digression: The "Throwers Elbow"

The "throwers elbow" is the result of overutilizing the muscle and tendon insertion on medial epicondyle of the humerus.

The explanation rests on the one hand on faulty throwing technique, on the other hand on anatomic circumstances.

In a *faulty throwing technique,* the arm is held too far lateral from the head, so that the arm and forearm form an angle of 90°, and the medial epicondyle points forward. Since the forward whipping motion is initiated by the arm, this part moves for-

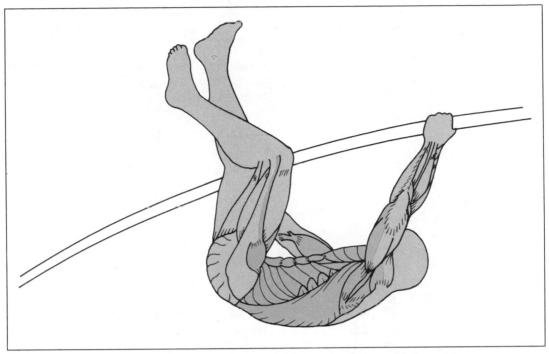

Fig 5–7.—The muscles participating in the upward roll phase of pole vaulting.

ward faster than the forearm, which results in the latter lagging behind and putting enormous leverage on the medial ligamentous bands and the medial epicondyle. The hinge joint of the elbow is now forced into an unphysiologic direction and thus placed under undue strain (Rau, 1969).

In addition to the increased liability to injury, this kind of faulty technique also places the arm, during the critical forward whipping motion of the arm, outside the optimal action line of the *pectoralis major m.*

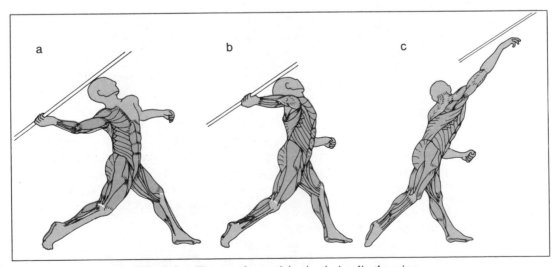

Fig 5–8.—The muscles participating in javelin throwing.

The decreased initial length, i.e., the decreased prestretching of the muscles leads to a decreased force of contraction. This is true, to a similar extent, for all muscles participating in the same forward whipping motion, and quite particularly for the *triceps brachii m.,* which has a controlling effect over the terminal extension of the arm.

The anatomic aspects of the "thrower's elbow" include the fact that at the very end of the throwing maneuver, there is an unusually forceful contraction of the flexors of the hand. Since all of these originate on the medial epicondyle, frequent throws can lead to local irritation in this area. Furthermore, the throwing arm is forcefully pronated from a previous supination. This leads to two actions on the *pronator teres m.,* which also originates on the medial epicondyle: (1) the explosive extension of the arm through the much stronger *triceps brachii m.* causes a stretching of the *pronator teres m.,* which normally acts as a flexor of the elbow, and (2) the pronator must contract to perform the pronation of the forearm. This twofold action can lead to muscle pain at the site of origin, microinjury, and subsequent scarring. Since, however, the pain occurs with whiplike speed at the time when the arm is thrown forward and the elbow is flexed (Rau, 1969), it can be assumed that the main stress on the medial epicondyle occurs at this time. It seems likely, therefore, that the irritation, characteristic of the "thrower's elbow," is due to excessive strain on the ligamentous system of the elbow joint and not to the simultaneously antagonistic strain on the *pronator teres m.*

Discus Throwing

The Muscles That Determine Performance (Fig 5–9)

LEG ACTION.—Similar to other throwing and tossing activities, great extensor strength of the hip, knee, and upper ankle joint is required for the final stretch acceleration, which must be transmitted from the legs and trunk to the throwing arm.

TRUNK ACTION.—Since the throw involves a twisting (prestretching) of the trunk followed by a rotation-extension, major contractile strength of the trunk rotators and extensors is required.

ARM ACTION.—In this respect, discus throwing is quite different from javelin throwing or shot put. While in the latter, the *triceps brachii m.* is of the greatest importance in performance, in the former it plays only a subsidiary role (it is active mainly in the recovery phase).

The primary throwing muscle in discus throwing is the *pectoralis major m.* It is strongly supported in the final throwing phase by the flexors of the arm (see p. 142) and the flexors and radial abductors of the hand.

Specific strength training of the *pectoralis major m.* can be accomplished by the use of barbells. This involves the raising of the arms, in the prone position from the lateral extension to the vertical position. (The thumbs should point upward.)

Shot Put

The Muscles that Determine Performance (Fig 5–10)

LEG AND TRUNK ACTION.—As in discus throwing.

ARM ACTION.—The *pectoralis major m.,* the short head of the *biceps brachii m.,* the *coracobrachialis m.* and the anterior portion of the *deltoid m.* bring the arm from the lateral position forward and, in doing so, support the accelerating anterior extension of the forearm by the *triceps brachii m.*

The final push by the wrist and fingers is accomplished by the respective flexors (see p. 142).

SPECIFIC TRAINING.—Push-ups with the hands turned inward (similar to the angle of

Olympic champion
Wolkermann at a perfectly
braced javelin throw.

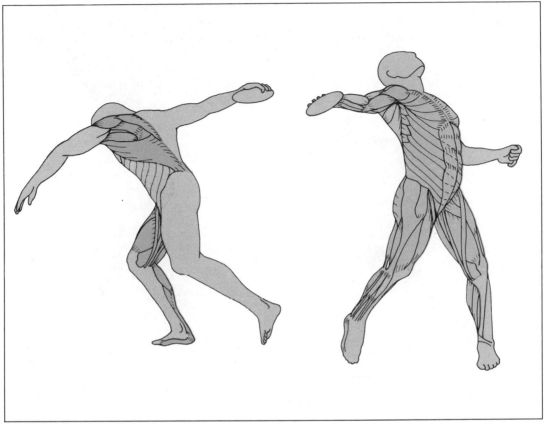

Fig 5–9.—The muscles participating in discus throwing, from the start of the throw to the extension of the trunk.

push and to the final position of the hands), and a hand clapping performed at the top of the push-up.

The Hammer Throw

The importance of muscle strength for the hammer thrower becomes evident when we realize that the thrower must impart to the hammer a flying force of 250 kp at the moment of release. This requires extremely strong leg, hip, and back extensor muscles and, also, powerful hand and finger flexors. Similarly, there have to be extremely strong muscles linking the arm to the shoulder joint (see p. 76) and, also, equally strong arm raisers (see p. 137) to accomplish the throw (Fig 5–11).

SPECIFIC TRAINING EXERCISES.—Composite exercises that serve all muscular components include "jerking" weights, throwing weights over the head backward, and swinging a colleague around and around parallel to the ground.

Swimming

In developing specific strength for swimmers, two principles must be observed: (1) only those muscles must be strengthened that serve the forward motion and stability of the trunk, (2) developing endurance is more important than maximal strength. The development of "show" muscles and too large muscles, not only creates an unfavorable specific gravity and reduces buoyancy,

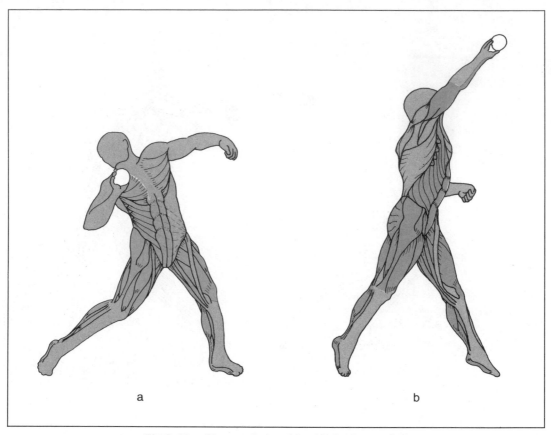

Fig 5–10.—The muscles participating in shot putting.

but the larger the muscle, the longer the distance over which oxygen and nutrient substances have to diffuse to reach the muscle cells.

The Breast Stroke

The Muscles That Determine Performance For Arm Action (Fig 5–12)

THE PULLING PHASE.—(See p. 138). Arm-lowering muscles, flexors of the arm, and flexors and ulnar abductors of the hand.

THE PUSHING PHASE.—Also the flexors and adductors of the arm (see p. 141) and the flexors and radial abductors of the hand.

THE RECOVERY PHASE.—Extensors and raisers of the arm (see p. 136).

Since in the breast stroke, as indeed, in all other forms of swimming, all muscles of the shoulder joint and all flexors and extensors of the arm and hand must participate, it is not surprising that the untrained swimmer very quickly develops a "heaviness" of the arms.

For Leg Action
Kick: Hip and Knee Flexors

SCISSOR KICK.—In the first phase of this maneuver, there is an internal rotation of the thigh (see p. 146) and an external rotation of the leg (biceps femoris m.) in which the foot is extended dorsally. In the second phase, there is an increasing hip extension, knee extension, and plantar flexion of the foot.

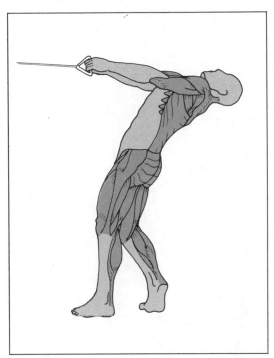

Fig 5–11.—The muscles participating in the final phase of the hammer throw.

To accomplish the up and down bobbing motion of the trunk, the swimmer requires adequately conditioned abdominal and back muscles.

DIGRESSION: THE SWIMMER'S KNEE.—The circular movement of the leg at the knee joint can lead to an overtaxing of the medial ligaments and tendons and thus to irritation of these structures.

This can be prevented by (1) strengthening the stabilizing muscles of the knee and (2) avoiding the premature extension of the knee during the kick.

Free-Style Swimming (Crawl)

The Muscles That Determine Performance For Arm Action (Fig 5–13):

THE PULLING PHASE.—The muscles forcibly lowering the arm (see p. 138) the flexors of the arm (see p. 138) and the flexors of the hand (see p. 142).

THE PUSHING PHASE.—Muscles lowering and extending the arm, flexors of the hand.

RECOVERY PHASE.—Primarily the *deltoid m.*

For Leg Action

DOWNSTROKE.—Flexors of the hip (see p. 143).

UPSTROKE.—Extensors of the hip.

Butterfly Stroke

The butterfly stroke is closely related to the crawl except that here, both the arm action and the leg action is performed by both sides simultaneously. The same muscles are being used.

Since the arm and leg action does not alternate from side to side, pronounced trunk movements become necessary (the so-called butterfly flutter). This requires increased mobility of the spinal column and well-developed abdominal and back muscles.

The Back Crawl

The maneuver is similar to the regular crawl but in a different body position. In the recovery phase, the anterior portion of the deltoid m. is particularly utilized. Leg action is also performed by the same muscles (Fig 5–14).

The great similarity in the muscles used and in the need for endurance in the crawl, back crawl, and butterfly stroke explains why a swimmer can perform extremely well in more than one type of swimming.

Fancy and High Diving

The Muscles That Determine Performance

For the jumping-off maneuver, the jumper needs strong hip extensors, knee extensors, and particularly strong foot flexors. To achieve sufficient hip flexion in the twist-turn jumps (Fig 5–15), the jumper needs extremely well-developed abdominal muscles (see p. 61) and flexor muscles of

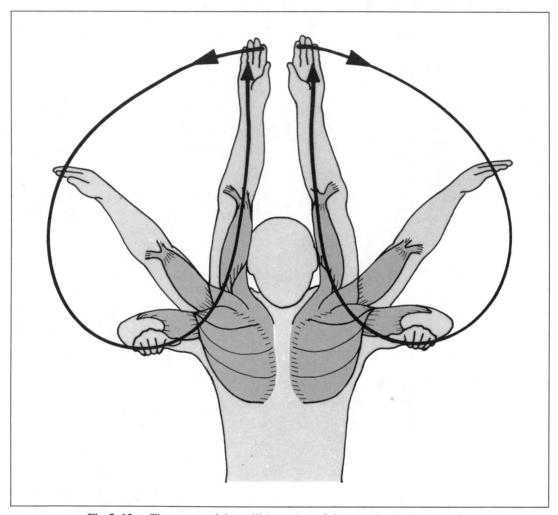

Fig 5–12.—The course of the pulling motion of the arms in the breast stroke.

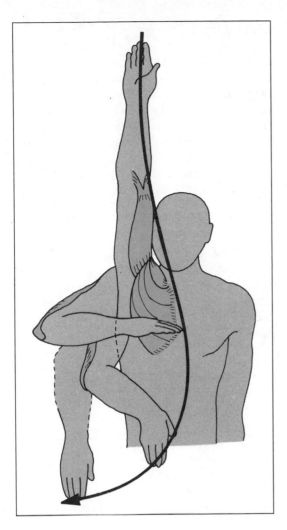

Fig 5–13.—The course of the pulling motion of the arm in free-style swimming.

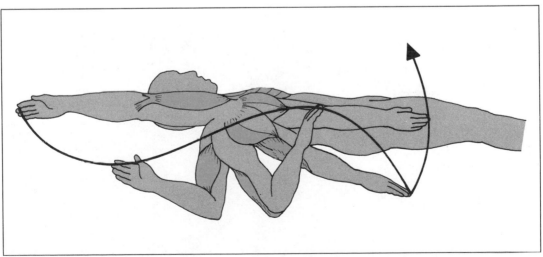

Fig 5–14.—The course of the pulling motion of the arms in the back crawl.

Fig 5–15.—The muscles participating in the jack-knife jump.

the hip (see p. 143), as well as extremely elastic ischiocrural muscles (see p. 106). To extend the trunk during the handstand, after the turns and on entry into the water, strong hip and back extensor muscles are required.

The handstand jumps also require strongly developed muscles to hold the arms in vertical extension, to extend the elbow joint, and to flex the wrist (see p. 142).

Water Polo

Water polo requires not only the development of the muscles needed for swimming, but also the training of the throwing muscles (see p. 161). Treading water requires strong and sustained effort of the adductor muscles and of the hip and knee extensors.

Huda in a pike position from
the tower.

Gymnastics

It can be stated generally that the majority of the maneuvers performed on the different gymnastic equipment can be divided into support, hand, or jump exercises. The complexity of many of the exercises requires a detailed analysis of the components that determine the performance.

Since it is not possible to analyze all the gymnastic exercises in this volume on the basis of their anatomic substrate, only some characteristic examples for each particular piece of equipment will be studied according to their dynamic requirements and anatomic distinctions. Many of these can be applied to a variety of other exercises, which will not be discussed.

As an introduction, it can be said that the *triceps brachii m.* is important in general support exercises, that the flexors of the hand are important in hang exercises, and that the extensors of the hip and knee, the flexors of the ankle, the extensors of the arm, and the flexors of the hand are important in the jumping exercises. In some of the hang exercises, e.g., the giant wheel on the horizontal bar or on the still rings, the flexors of the hand and fingers must overcome a multiple of the total body weight.

The Horizontal Bar

Since in modern practice, the exercises on the horizontal bar consist almost entirely of swings (except for momentary support phases), it is the strengthening of the flexors of the hand and fingers that deserves particular attention. This conditioning has to be specific for this equipment and has to be done in such a fashion that the other components of the swing sequence, i.e., upper extremity and trunk, also receive their due share. Thus, the flexors and extensors of the elbow, the stabilizers of the shoulder, and of the shoulder girdle must also be appropriately and adequately strengthened (see pp. 73 and 76).

The Support Kip Exercise

The Muscles That Determine Performance

On the horizontal bar, the still rings, and the parallel bars, the kip is entered from a support phase, through a hang phase and back into a support phase again. In the hang phase, abdominal muscles (which stabilize the position of the pelvis) and the flexors of the hips (see p. 143) prevent the legs from touching the ground. At the moment of the kip maneuver, there is an explosive extension of the hips (*gluteus maximus* and *ischiocrural* muscles) and the simultaneous approximation of the arms to the trunk. By fixing the upper extremity, the arm-lowering muscles pull the trunk upward (reversal of the fixed and mobile points) and support the reassumption of the starting support phase.

The Parallel Bars

The parallel bars are primarily a support equipment and therefore require particular conditioning of the arm extensors and of the muscles of the shoulder girdle.

The Handstand

This exercise is performed in various forms on all gymnastic equipment, either as specific exercise, or as a quick transient phase of a more complex move.

The Muscles That Determine Performance (Fig 5–16)

The handstand requires a substantial support strength of the arms *(triceps brachii m.)* and a fixation of the upper extremity in overhead elevation (see p. 137). In addition, the entire body must be held straight and in balance by the isometric contraction of all the flexors and extensors.

The Still Rings

The still rings are both a swing equipment, similar in many to the horizontal bar, and also a support equipment. It has the peculiar feature that the rings can move in any direction. This allows a large number of different power maneuvers that are quire unique for this equipment.

The Iron Cross (Fig 5–17)

The iron cross requires an extraordinary amount of strength from the flexors of the hand, the extensors of the arm, the adductors of the arm (primarily the *pectoralis major m.* and *latissimus dorsi m.,*) and the stabilizing muscles of the shoulder (see p. 76).

The Backward Hanging Scale (Fig 5–18)

This exercise requires great strength from muscles, which extend the arms forward, particularly the *pectoralis major m.* and the *biceps brachii m.* In addition, an extreme strength of the trunk, particularly of the *erector spinae* and *latissimus dorsi m.,* is a prerequisite. The *latissimus dorsi m.* pulls the posterior rim of the pelvis toward the arms through its insertion, via the lumbodorsal aponeurosis, and it is primarily this action that makes the backward hanging scale possible. The *gluteus maximus m.* and

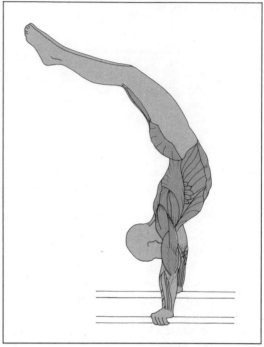

Fig 5–16.—The muscles participating in the handstand on the parallel bars.

the *ischiocrural m.* support the extension of the hips.

The ''L''' Support Position

This exercise depends on the support musculature of the upper extremity. In addition, great strength is demanded of the flexors of the hip (see p. 143) and of the abdominal muscles, which fixate the position of the pelvis (see Fig 4–6).

The Pommel Horse

The pommel horse is *the* support equipment in gymnastics. To sustain the support position for the required period of time, it is essential that the extensors of the arm and the flexors of the hand be conditioned and developed for both peak performance and endurance. In addition, some exercises, e.g., the scissors, require very strong abductors of the leg *(gluteus medius m.)* (see p. 205 for specific strength training).

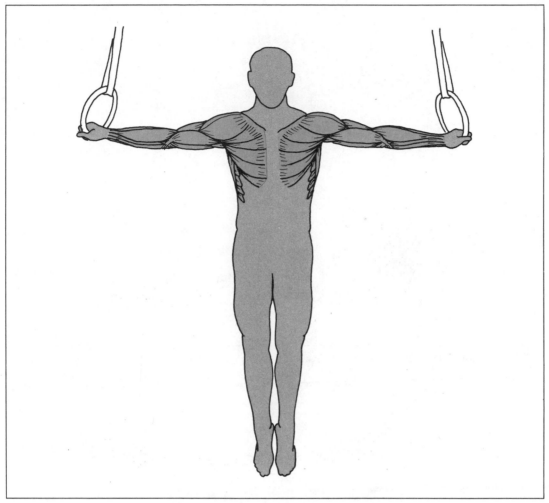

Fig 5–17.—The muscles participating in the "iron cross" on the rings (the trunk and the arms are extended).

Floor Exercises

The floor exercises consist of a combination of rolls, summersaults, kips, jumps, straddles, and static elements like stands and arabesques. All of these require that the hips be very flexible, and all muscles acting on the hips be strong. The free jumps (salti, etc.) require strong jumping muscles (see p. 156), and the cartwheels require powerful extenders and pushing muscles (particularly the superficial and deep flexors of the fingers).

To perform a front scale *(arabesque)* (Fig 5–19), strong hip stabilizers, particularly abductors and adductors on the side of the support leg, are required (see pp. 103 and 105). This side also requires a balancing activity on behalf of the positional muscles of the foot (see p. 118) and of the plantar flexors. On the side of the swing leg, the hip extenders are important.

Finally, fixation of trunk in a straight line is accomplished by the *erector spinae muscles.*

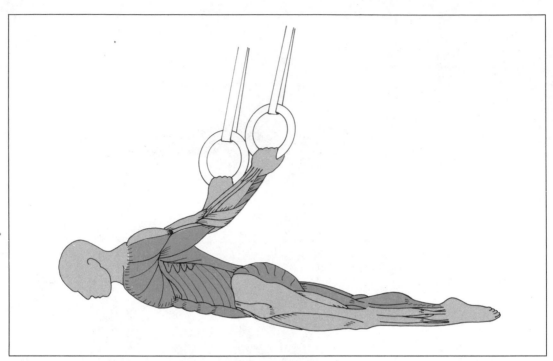

Fig 5–18.—The muscles participating in the backward hanging scale on the rings.

Fig 5–19.—The muscles participating in the front scale (arabesque).

Vaulting

Vaulting requires the same leg muscles, as those already discussed for other gymnastic jumps (see p. 156). In addition, because of the preflight and repulsion phases of this exercise, the same support and pushing-off muscles of the arms are required, as in the floor exercises. In fact, there is a greater need for good arm muscles, because the speed of hurdling and of the preflight, does put increased demands on the upper extremity.

The Uneven Bars

Gymnastic exercises on the uneven bars are a combination of the hang maneuvers on the horizontal bar and the support exercises on the parallel bars. The muscles needed for both must therefore be developed accordingly. The muscles primarily involved are the extensors of the arm, the flexors of the hand as well as the flexors, the extensors, and the adductors and abductors of the hip.

Fig 5–20.—The muscles participating in rising from a deep squat.

The Balance Beam

The balance beam exercises are the most demanding and most difficult gymnastics in the female repertory.

The exercises are basically modifications of those performed in the floor exercises. These include jumps (forward and backward standing cartwheels), support exercises, and acrobatics, requiring an active straddle and longitudinal split ability on the part of the gymnast.

The balance beam exercises require not only superb coordination and a high degree of mobility, but also an impressive amount of truncal and extremity strength.

Since the standing and support base is very narrow, the positional muscles of the foot are very important in improving the ability of the gymnast to maintain his or her equilibrium. Specific strengthening exercises: the positional muscles of the foot can be strengthened by walking on tip toes and on the inner and outer edge of the foot (see p. 118).

Weight Lifting

The Muscles That Determine Performance (Fig 5–20)

ARM MUSCLES.—In the pulling phase, a strong grip is required (*flexor digitorum su-perficialis and profundus, flexor carpi radialis and ulnaris m.*) as well as maximally developed arm flexors and abductors.

In the "jerk," an important role is played by the *extensor digitorum communis, the extensor carpi radialis and ulnaris, longus and brevis m.*

The extension of the arm needs a strong *triceps brachii m.*, while the fixation of the arm in elevation is accomplished by the corresponding stabilizing muscles (see p. 137).

TRUNK MUSCLES.—The strength of the extensors of the trunk play a decisive role in the pulling phase and in holding the weight overhead. In this latter step, these muscles, together with the abdominal muscles, assure the stabilization of the trunk as a support column.

LEG MUSCLES.—Since the weight must not only be accelerated vertically by the hip extensors, the knee extensors, and the ankle flexors, but must also be stabilized in the overhead position, the entire musculature of the lower extremity assumes a most important role as lifters, and also as support column. The extreme strain on the knee joint in the deep kneebend position has already been discussed (see p. 14).

The Martial Arts

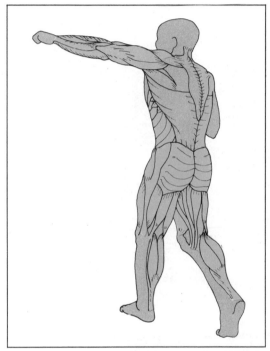

Fig 5–21.—The muscles participating in a "straight left."

Boxing

The performance of a boxer depends on strength, speed, and endurance, and their subdivisions, explosive power, speed endurance, and strength endurance.

The Muscles That Determine Performance

ARM MUSCLES.—In straight blows (Fig 5–21), the *triceps brachii m.* is most important. In uppercuts and roundhouse swings, the flexors of the arm and the muscles that move the arm anteriorly are significant par-ticipants. The *deltoid m.* and the flexors of the arm are responsible for keeping the lead arm up.

TRUNK MUSCLES.—The boxer needs strong back and abdominal muscles, both to support his own blows and also to resist the blows of the opponent. A significant conditioning of these muscles is also an absolute requirement for bobbing, weaving, and ducking.

LEG MUSCLES.—In the "follow-through" after a blow, the extensors of the hip, extensors of the knee, and flexors of the ankle are of great importance.

Wrestling

The wrestler is the power and endurance athlete par excellence, since he has to activate maximal and explosive forces of the entire locomotor system in rapid sequence.

The Muscles That Determine Performance

ARM MUSCLES.—In the wrestler, the muscles of the upper extremity must be extraordinarily well developed for every type of activity. The flexors of the arm, the flexors of the fingers, and the adductors are of particular importance for the various holds.

TRUNK MUSCLES.—In the general strengthening of the trunk muscles, particular attention must be paid to the extensors and rotators of the trunk (see p. 135). Some wrestling moves can be performed only af-

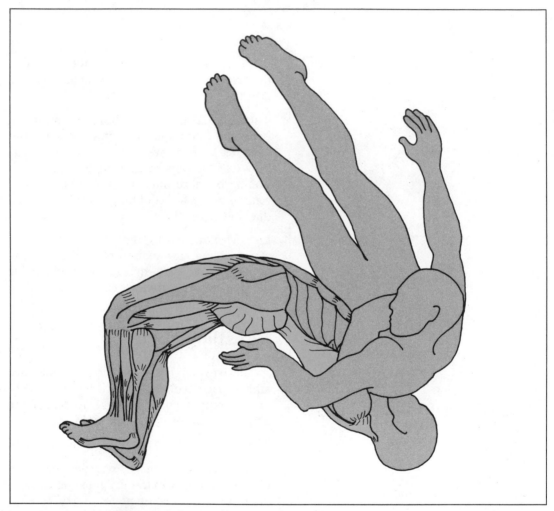

Fig 5–22.—The muscles participating in a reversal throw.

ter specific strength training, e.g., bridging exercises for the supine reversal moves (Fig 5–22).

LEG MUSCLES.—To perform the aggressive and defensive maneuvers, the wrestler needs strong hip extensors, hip abductors and adductors, as well as knee extensors and ankle flexors.

Fencing

The Muscles That Determine Performance (Fig 5–23)

ARM MUSCLES.—All flexors and extensors of the hand are used, as a group, in the positioning of the weapon arm and in the stabilization of the wrist. For forward lunges, the abductors of the arm and the extensors of the arm come into play. In "invitation" moves (e.g., quart invite), rotation of the forearm along the longitudinal axis (pronation and supination, see p. 141), as well as simultaneous internal rotation *(subscapularis, pectoralis major m.)* and external rotation *(infraspinatus m.)* of the arm at the shoulder are required. Strengthening all these muscles, but particularly the abductors of the arm, is critical.

TRUNK MUSCLES.—The trunk muscles, on the one hand, support the attack maneuvers (primarily the abdominal muscles and the flexors of the hip), while on the other hand, they make rapid evasive moves possible during the defensive maneuvers. A well-developed trunk musculature is a fundamental prerequisite for the fencer.

LEG MUSCLES.—The explosive forward leap is the most characteristic leg action of the fencer, since every attack is based on it. Since "you are as good a fencer as you can leap forward" is a basic principle in fencing, the hip flexors, the extensors of the hip and knee, and ankle flexors play a dominant role (see pp. 103–105).

The acceleration that the fencer has to perform in the moment of the forward lunge can be compared in intensity with the start of a track and field event for the sprinter. A good start is half the battle, but for the fencer, it is *the* battle, since the distance he or she has to overcome is so much less.

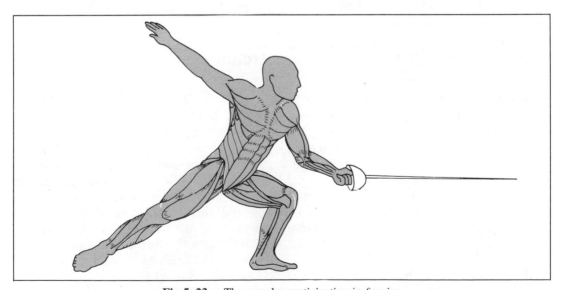

Fig 5–23.—The muscles participating in fencing.

Target Shooting

Target shooting with the rifle and archery will be analyzed according to their anatomic substrate, as examples for the different shooting competitions.

Rifle Shooting (Standing)

For the rifle shooter, the most important feature is not the maximal strength of the participating muscles, but their ability to sustain the required power over several hours. For this, dry runs (without ammunition) are usually sufficient. Optimal firing stance, positioning the rifle correctly, holding the rifle steady, learning and improving the pulling of the trigger, and coordinating the correct holding of the rifle with the correct trigger pull, can all be learned in this way. In addition, general physical conditioning is important to assure the required general muscular strength and agility.

The Participating Muscles

TRUNK AND LEG MUSCLES.—While aiming the rifle, the body is in a labile equilibrium, which results in a necessary tightening of the entire positional muscular system (Fig 5–24). The backward slant and simultaneous rotation of the trunk requires the bracing of the trunk and of the hip joints. The former is accomplished by muscles, the latter by the ligaments of the spinal column and of the hips *(iliofemoral ligament)* (see pp. 55 and 97).

The lateral bend of the trunk is stabilized mostly by the abdominal muscles, *the erec-*

tor spinae, the quadratus lumborum and the iliopsoas m. The fixation of the pelvis is also contributed to by the abductors and adductors of the thigh (see p. 103), and the flexors and extensors of the hip.

The maintenance of equilibrium, critically important to the rifleman, is accomplished by the *triceps surae and tibialis anterior m.*, acting on the upper ankle joint.

ARM MUSCLES.—By supporting the left elbow on the left hip bone, much of the burden is lifted off the shoulder muscles. The rifle is stabilized with relatively little effort by the flexors of the arm. This task is made easier by the narrow angle between the arm and forearm. The trigger arm is held in abduction at the shoulder by the *deltoid m.* Steadiness and endurance of this muscle are important in taking accurate aim.

Archery

The Muscles That Determine Performance

ARM MUSCLES.—During the draw, the aim and the release, the arm and hand must hold the bow steady. This requires both strength and endurance from the muscles of the bow hand, particularly the extensors of the arm and the arm raisers. Since the wrist of the bow arm must be extended so that the pressure caused by drawing the bowstring is transmitted to the forearm in a straight line, the wrist extensors and flexors also require considerable strength.

The abductors of the arm must be as

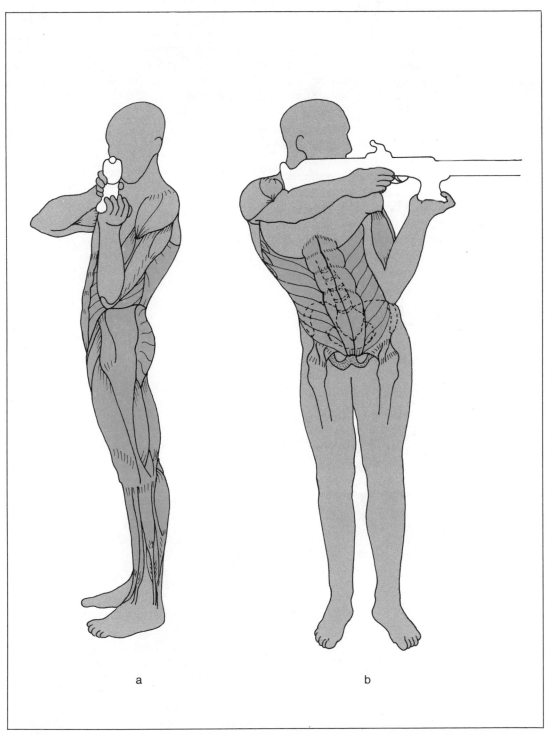

Fig 5–24.—The muscles and ligaments participating in aiming a rifle in the standing position (after *Jurjew*).

a
b

strong as possible, to allow a wide choice of bows and stabilizers. A heavy bow has considerable advantages for the archer (it "sits" better in the hand, does not move as readily when the drawstring is released, and is less affected by wind), but also requires increased strength in the muscles described above.

Since the physical condition of the archer must correspond to the weight of the bow and of the stabilizers, to avoid premature fatigue of the bow arm and, consequently an inability to hold the bow steady, the adequate conditioning of the appropriate muscles is a prerequisite to achieve a better score.

The arm of the arrowhand is moved from an anterior to a lateral position *(infraspinatus, deltoid, teres minor m.)* at the same time that the elbow is flexed. These muscles, as well as the flexors of the fingers, must be strengthened by appropriate exercises (see p. 205).

TRUNK AND LEG MUSCLES.—Erect posture is a basic element of good archery. Conditioning of the trunk muscles and of the stabilizers of the hip *(adductors and abductors)* is essential to prevent an undesirable rotation of the trunk (see p. 105).

Watersports

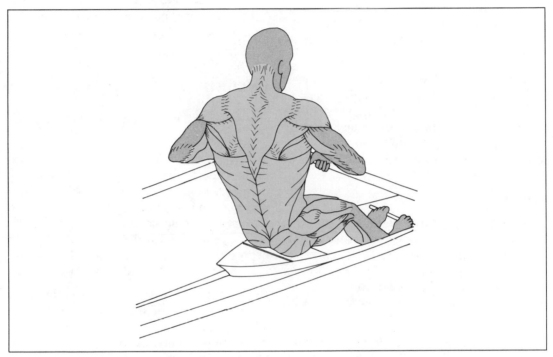

Fig 5–25.—The muscles participating in rowing.

Rowing

Rowing is a power *and* endurance sport that makes high demands on the arm, trunk, and leg muscles (Fig 5–25).

Arm Muscles.—Flexion is accomplished by the *biceps brachii, brachialis, and brachioradialis m.*, the retroversion and lowering of the arm are the function of the *pectoralis major, triceps brachii, latissimus dorsi, teres major, and subscapularis m.*

Trunk Muscles.—All abdominal muscles are used in bending the trunk forward. Bending the trunk backward (pulling phase) is the function of the *erector spinae m.*

Leg Muscles.—The pulling phase is characterized by an increasing extension of the hip joint and knee joint and a plantar flexion of the upper ankle joint.

Kayak (Paddling)

The Muscles that Determine Performance (Fig 5–26)

Arm Muscles.—The pulling arm is retroverted from a lateral position and simultaneously lowered (see "rowing"). The elbow is flexed and the wrist is flexed.

The pushing arm (pulling and pushing occur simultaneously) is increasingly extended by the *triceps brachii m.* and raised by the *deltoid m.*

TRUNK MUSCLES.—The powerful trunk rotators support the arm pull (see p. 135). The initial insertion of the blade and sharp tug occurs when the trunk is maximally prestretched and rotated. To stabilize the trunk, strongly conditioned abdominal and back muscles are essential.

LEG MUSCLES.—The leg muscles are much less actively involved in paddling than in rowing. They serve, however, to stabilize the body in the sitting position and to assist in the pulling and pushing motions of the arms. The most important contribution is made by the flexors of the hip.

Sailing

Modern, competitive sailing (regatta), requires considerable bodily performance. To be able to maintain the very demanding outrider position for two to four hours during a race, under favorable wind conditions, requires that the appropriate muscle groups be very highly conditioned.

The Muscles That Determine Performance

ARM MUSCLES.—To hold and raise the sails, to hold and take in the lines, and to hold the tiller in a leeward drift require that the sailor have strong arm flexors and retrovertors of the arm (*deltoid, subscapularis, teres major m.*).

TRUNK AND LEG MUSCLES.—To assume the outrider position and to return from this position makes the strengthening of the abdominal muscles and of the flexors of the hip *(rectus femoris, iliopsoas, tensor fasciae latae m.)* absolutely essential. To squat and to rise from a squat needs strong knee extensors. To hold the feet in the outrider shackle requires strong plantar flexors.

Since the muscles perform primarily holding actions, particular attention must be paid to the development of endurance.

Cycling

The Muscles That Determine Performance (Fig 5–27)

LEG MUSCLES.—The leg muscles are the primary performers in cycling. Depressing the pedal is accomplished by the hip extensors and, more importantly, by the knee extensors and ankle flexors. Raising the pedal is accomplished by the antagonists of the depressor muscles, i.e., hip flexors, the knee flexors, and the extensors of the ankle.

ARM MUSCLES.—In normal cycling, the extensor muscle of the arm serves to steer and to maintain the trunk in the optimal position. In the spurt, the flexors and the muscles that lower the arm (see p. 138) become important.

TRUNK MUSCLES.—To transmit the support work of the arms to the legs, a well-conditioned abdominal and back musculature is essential.

An impression from the Kieler
Week.

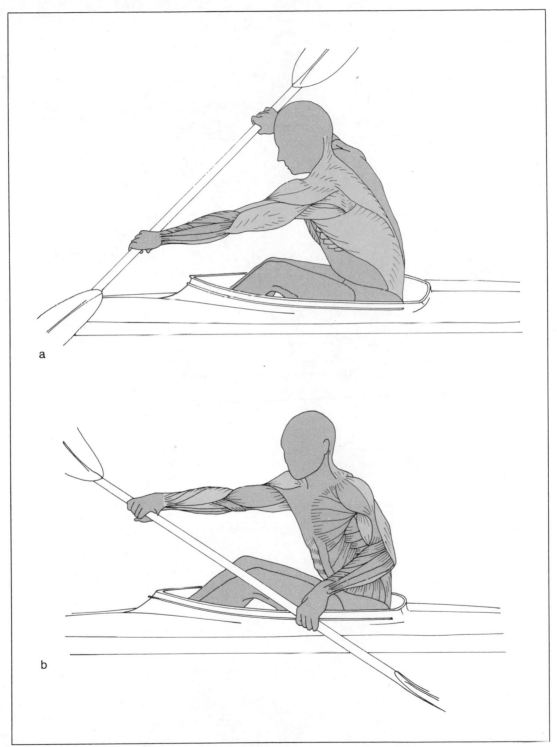

Fig 5–26.—The muscles participating in paddling a kayak.

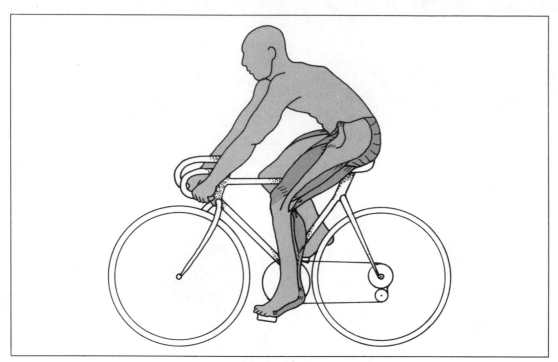

Fig 5–27.—The muscles participating in bicycle riding.

The Field and Court Sports

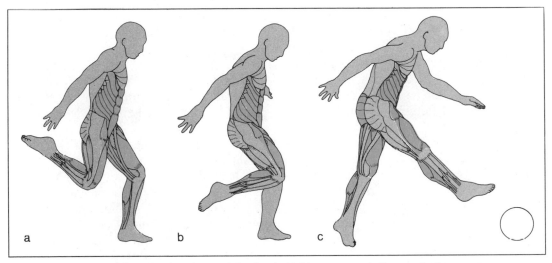

Fig 5–28.—The muscles participating in kicking a soccer ball towards the net.

To be effective in any of the field and court sports requires appropriate running and jumping muscles (see p. 156), in addition to certain specific muscle groups that are required by the particular technical demands of the individual sports. Only the latter, game-specific, locomotor activities will be analyzed in detail below.

Soccer

Kicking the ball toward the net requires an explosive extension of the knee joint and a flexion of the hip at the same time that the abdominal muscles contract. The activity of the kicking leg is supported, on the side of the standing leg, by an extension of the hip, of the knee, and plantar flexion of the ankle (Fig 5–28).

Field Hockey

The forceful and rapid handling of the stick requires that the hockey player have both strong arm flexor muscles and also strong arm abductors and adductors (see p. 141). The *pectoralis major m.* plays an important role in accomplishing a forceful shot. The flexion of the trunk of the player handling the ball requires well-conditioned extensor muscles of the back.

Handball

The modern game of handball places great demands on the athletic ability of the players. This makes a very pronounced conditioning of the entire trunk and extremity musculature a major prerequisite. To improve the throwing ability all the muscles described in the section on javelin throwing must be strengthened. The side arm throw, accomplished with the forearm without any wind-up, requires the training of the elbow flexors, wrist flexors, and finger flexors. Strong finger flexors are also important in guiding and aiming the ball.

Basketball

Basketball is a game that requires the ability to jump both with one foot and with both feet. It also requires sufficient arm extensor strength to throw the ball. The throws require more accuracy than force; nevertheless, the longer throws do require a certain amount of throwing strength.

Volleyball

Volleyball is a game that requires a combination of two-legged jumps with a throwing motion. This throwing motion has elements of a downward blow (smash), which requires the development of appropriate strength (see ''Javelin'').

Tennis

To obtain a solid grip, the tennis player needs strong finger flexors. To stabilize the wrist, he/she will also need well-developed wrist flexors and extensors.

The *triceps brachii m.* extends the arm for the backhand stroke and for serving. In the forehand stroke, the arm is brought forward by the *pectoralis major m.*, the anterior portion of the *deltoid m.*, and the *biceps brachii m.* The abduction and simultaneous retroversion in the backhand stroke is accomplished by the posterior portion of the *deltoid* and the *infraspinatus* and the *teres minor m.*

The power serve is accomplished by the same muscles as the javelin throw (see p. 161) and involve, above all others, the *pectoralis major* and *latissimus dorsi m.*

Digression: "Tennis Arm"

''Tennis arm,'' or ''tennis elbow,'' refers to a strain that is manifested by signs of irritation at the external epicondyle of the humerus.

The symptoms occur primarily after a backhand stroke, since in this maneuver, stretch of the wrist and elbow is followed by a terminal supination of the forearm. Since all extensors and supinators originate on the external epicondyle of the humerus, it is easy to see how technically poor strokes (uncontrolled extension and supination due to excessive force), or playing with a metal racket can lead to local damage through excessive stress.

The ''tennis elbow'' is naturally not limited to the external epicondyle. In the forehand stroke and serve, the medial epicondyle undergoes very similar stresses.

Winter Sports

Alpine Skiing

The Muscles That Determine Performance

LEG MUSCLES.—The dynamics of Alpine skiing require an extraordinary strength and endurance from the extensors of the hip and knee to maintain the "downhill crouch" and the rapid, powerful push in the turns. The strength of the extensors (*tibialis anterior m.*) and flexors (*triceps surae m.*) is also important.

In studying the "downhill crouch," it is important to point out a biomechanical and infrastructural peculiarity of the *quadriceps femoris m.* and its components (see p. 110). Studies in the wind tunnel suggested that the so-called "low egg" position (Fig 5–29, a), which depends largely on the *rectus femoris m.* having rapid, powerful action but limited endurance, had advantages. Actual experience in world competition showed that this was not true and that the "high egg" (Fig 5–29, b) was preferable. This position relies mostly on the *vastus femoris m.*, which are designed, quite particularly for isometric holding action. The illustrations clearly show the difference in the extensor axis of the knee between the two positions.

In the "low egg" position (deep knee bend) the *rectus femoris m.* and the *vastus intermedius m.* lie over the extensor axis of the knee; the *median and lateral vastus m.* lie below it. Consequently, the *rectus femoris m.*, supported only by the *vastus intermedius m.*, must perform a sustained

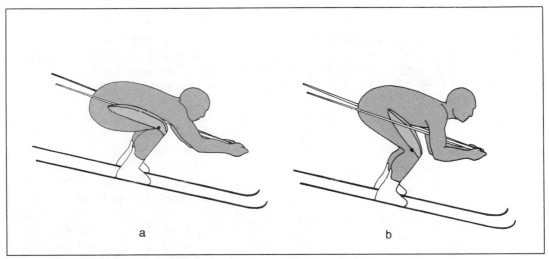

Fig 5–29.—Diagrammatic representation of the performance capabilities of the two components of the quadriceps femoris m. and their relationship to the rotational axis of the knee joint: **(a)** low egg shape and **(b)** high egg shape.

holding action for which it was not designed and which quickly overtaxes it.

In the "high egg" position, the *rectus femoris m.* and *all three vastus m.* lie above the extensor axis. Consequently, all components, including the *vastus femoris m.*, participate in the sustained holding action for which they are eminently suitable.

TRUNK MUSCLES.—The Alpine skier needs strong back muscles to maintain the aerodynamically favorable downhill crouch. Strong abdominal muscles are also needed, particularly for the twists and turns of the downhill slalom.

ARM MUSCLES.—For the placement of the poles and for the recovery from falls, strong extensor muscles of the arm are required.

Cross-Country Skiing

The Muscles That Determine Performance

LEG MUSCLES.—To have a powerful push, the leading leg needs strong hip extensors, knee extensors, and plantar flexors. To carry the rear leg forward, strong hip flexors are needed.

ARM MUSCLES.—The pushing arm needs the *triceps brachii m.* and the muscles lowering the arm (see p. 138). The swing arm is moved primarily by the *deltoid m.*

TRUNK MUSCLES.—The abdominal and back muscles provide the stability of the trunk, and support the leg extensors.

Speed Skating

The Muscles That Determine Performance

LEG MUSCLES.—Extension of the legs is accomplished in the same way as in running. Increased strength is required, however, of the adductors (particularly in the turns) and of the abductors of the hip joint.

TRUNK MUSCLES.—The sustained, marked forward flexion of the trunk is accomplished primarily by a contraction of the *erector spinae m.* The abdominal muscles, as an-

tagonists of the pelvic stabilizers, should also not be ignored.

Figure Skating

In figure skating, strong positive (jumping) and negative (landing) dynamic power is required. For this reason, both the jumping and equalizing muscles, as well as the positional muscles of the foot, must be markedly strengthened (see pp. 118 and 156).

The male partner in pair skating requires abdominal, back, arm lift, and arm extensor muscles nearly as strong as those of a weight lifter.

Bobsledding

Bobsledding consists of a starting phase and of a ride toward the finish line, which is independent of bodily strength and requires only technical skill in steering.

The muscles needed in the starting phase are the same as those used by the sprinter (see p. 155), although with an increased need for strong extensors of the arm and stabilizers of the trunk.

Sledding

To obtain the longest possible acceleration path in the initial "get set" phase at the start, the sledder requires a highly flexible spinal column and strong hip flexors to allow him or her to lean far forward.

In the "moving-off" phase of the start, it is the strength of the arm flexors, of the arm-lowering muscles (see p. 138) and of the retrovertors that matter.

Finally, in the "turn-over" phase, the starting speed can be increased by pushing off with the hands (penguin move). In the initial pulling phase, the flexor muscles of the arm determine the performance; in the pushing phase, it is the extensor musculature that supports the activity.

During the ride, superbly developed neck muscles *(sternocleidomastoid m.)* and abdominal muscles are essential, since the

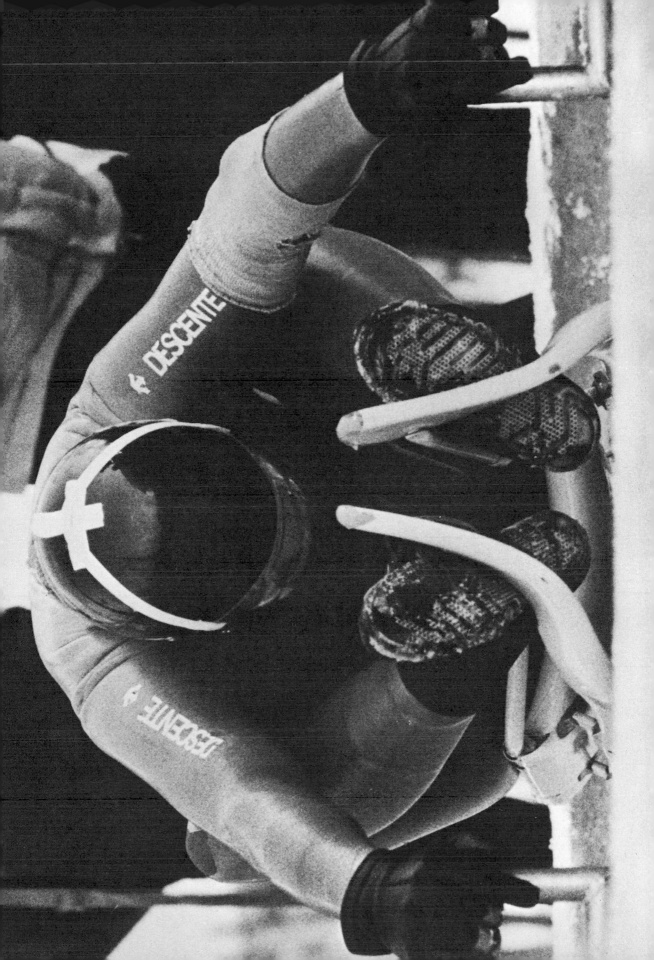

Concentrated energy at the
gate: Winkler at the Olympic
Games, Lake Placid, NY.

head must be slightly elevated throughout the ride to see the glide path. The demands of aerodynamics (with increasing speed the air resistance increases as a square function) and the necessity for precision steering are antagonistic and place a great burden on the neck and abdominal muscles.

In addition, to properly stabilize the sledder in the sitting position, the adductors of the "lead hand" (particularly the *pectoralis major m.),* the flexors of the "strap hand," and the adductors of the hip must be powerfully conditioned.

Ski Jumping

In ski jumping, there is a combination of an extreme positive (jumping off) and an extreme negative (landing) dynamic stresses on the hip extensors, the knee extensors, and the plantar flexors of the foot.

To stabilize the body in an optimal flying posture, well-developed abdominal and back muscles are essential.

6 Functional-Anatomic Strength Training

Introduction

This is an introduction to a goal-oriented, functional-anatomic strength-training program, involving all the muscle groups that were included in the section entitled "Analysis of Simple Locomotor Functions—Advice Concerning Dynamic and Static Exercises."

It is not the purpose of this section to describe every possible exercise for any given motion. It is the purpose, however, to give the nonspecialist simple and modest advice in how to correct muscular weakness in a convenient way. For this reason, only one single exercise is offered for each motion. Additional exercises can be devised by the reader, or found in the volume *"Optimal Training"* by the author.

What Must be Considered in Strength Training?

- To avoid muscle pain (charley horse) in the initial phases of strength training, it is better to perform an exercise more frequently and with lesser loads.
- Whether right or left handed, do not forget the other side!
- Strengthen not only the agonists, but the antagonists as well.
- Muscles can be trained singly and in combination.
- No strength training without warm-up period.
- Strength training has to be specific for the requirements of the individual sport or athletic event. This means: maximal strength training, when maximal strength is required; endurance training (submaximal load, but higher total number of performances), when this form of power is required.
- Training has to be done so that the natural dynamics of every athletic maneuver are closely imitated, and the angle of attack remains the same.
- Strength can be acquired through dynamic training (the muscle is shortened) or static training (the muscle is only maximally tightened).
- In dynamic training for great strength, a repetition of the exercise eight times is optimal (weights have to be chosen so that this number can be reached). In static training, isometric contraction should be sustained for six to ten seconds (this equals one series).
- Two to five series should be performed during every training session, depending on the level of training.
- Static training has no effect on circulation. It purely increases the diameter of the muscle cell and so increases muscle strength.
- Since static training involves great straining, it is not suitable, without further study, for people with cardiac problems or high blood pressure.

A List of Exercises for Simple Locomotor Processes

(see p. 129 et seq.)

Simple Trunk Movements

Flexion Forward

Dynamic exercise: Bend the trunk forward, with the legs straight.

Static exercise: Tighten the abdominal muscles (panting respiration).

Flexing the Trunk Backward

Dynamic exercise: Raise the trunk from the supine position. Legs fixed.

Static exercise: Tighten the back muscles while stretching.

Lateral Flexion of the Trunk

Dynamic exercise: Bend the trunk sideways (the legs are spread and fixed).

Static exercise: Bend the trunk sideways and remain in the end position.

Rotation of the Trunk

Dynamic exercise: Rotating the trunk left and right under additional weight (sandbag, horizontal bar, or weight-lifting bar).

Static exercise: Stand in the doorway with the arms extended and the legs fixed. Try to rotate against the resistance (alternating right and left).

Simple Movements of the Upper Extremity

Lifting the Arm From the Anatomic Position to the Horizontal Anterior and Overhead Positions

Dynamic exercise: Perform the movement with barbells.

Static exercise: Both arms forward; the palm of one hand rests on the back of the other hand (use different forward elevations); now push the hands against each other, the upper hand downward and the lower hand upward. This will strengthen the arm-lowering and anteverting muscles, respectively.

Fixation of the Arms in the Overhead Position

Dynamic exercise: Walk on your hands.

Static exercise: Same as just above, but with the arms in the overhead position.

Lowering the Arms From the Elevated Position

Dynamic exercise: Pulling exercises of any type (pulley, coil, etc.). In the supine position, raising the extended arms from the overhead position to the anterior position using barbells.

Static exercise: As above.

Retroversion of the Arms

Dynamic exercise: Pulling exercises from the standing (anatomic) position. Lifting barbells backward in the prone position.

Static exercise: Stand with your back against the wall; arms down, palms against the wall; push against the resistance.

Abduction of the Arms

Dynamic exercise: Lift barbells from the anatomic position to the lateral horizontal position.

Static exercise: Stand in the door, in the anatomic position. Push with both arms laterally against the door frame.

Retroversion From the Lateral Position

Dynamic exercise: In the prone position on a bench, lift barbells from the dependent (forward) position to the side and beyond.

Static exercise: Stand with your back to the wall; arms extended laterally with the palms against the wall. Push against resistance.

Anteversion From the Lateral Position

Dynamic exercise: While lying supine on a bench, lift barbells from the lateral to the vertical position (up).

Static exercise: Face the wall with the arms extended laterally, palms against the wall. Push against resistance.

Adduction of the Arms

Dynamic exercise: Grasp the rings with the arms extended laterally. Jump into the support position, bringing the arms down.

Pulley and weight exercise can be substituted.

Static exercise: Sit on a chair with the arms hanging at the side. Push against the chair from the sides.

Internal Rotation of the Arms

Dynamic exercise: Flex elbows to 90°. Sit across the table with a colleague. Push against the colleague's arm (arm wrestling).

Static exercise: Kneel down facing a chair, grab a chair leg with one hand, try to twist it against resistance (elbow should be bent at 90°).

External Rotation of the Arms

Dynamic exercise: As above, but with the back of the hand against the colleague; push outward.

Static exercise: As above, but in the opposite direction.

Flexing the Arm at the Elbow

Dynamic exercise: Chin-ups.

Static exercise: Sit at a table, place one palm against the table from below and one from above. Try to flex the lower arm and extend the upper arm (see "Extensor Muscles of the Arm").

Extending the Arm at the Elbow

Dynamic exercise: Push-ups, handstand on the parallel bars, raising weights in the supine position.

Static exercise: As above.

Rotation of the Elbow

Dynamic exercise: Rotation of a metal exercise rod, alternating pronation and supination.

Static exercise: See exercise for internal rotation, but performs with arms extended.

Flexion of the Wrist

Dynamic exercise: Pushing off in push-ups.

Static exercise: Hold hands with palms together like in prayer, push hands against each other.

Extension of the Wrist

Dynamic exercise: Twist a weight-lifting bar, wind a string with a weight at the end around a bar.

Static exercise: Place the hands on top of each other, push with the lower hand against the upper one.

Simple Movements of the Lower Extremity

Flexion of the Hip

Dynamic exercise: Raise legs while hanging from the horizontal bar. Flex hip forward, as in the abdominal muscle exercise.

Static exercise: Sit on a chair and cross your legs; push with the lower thigh against the upper thigh.

Extension of the Hip

Dynamic exercise: Squat-jumps, knee bends, and kip maneuvers of all types.

Static exercise: As above, but push with the upper part of the thigh.

Overextending the Hip

Dynamic exercise: Prone position arms extended and fixed; raise the thighs off the floor.

Static exercise: Back against the wall; the two feet in front of each other; push with the heel of the backward leg against the wall.

Abducting the Thigh

Dynamic exercise: Lying on the side; arms fixed; raise upper part of the thigh into the air.

Static exercise: Stand in the middle of the door frame and move one leg laterally until it touches the door frame with the outer edge of the foot; push against the resistance. (Work with both legs in sequence.)

Adduction of the Thigh

Dynamic exercise: Pulley and weight exercise, with the weight laterally; all jumps and sprints with frequent direction changes.

Static exercise: Sit with legs facing a chair. Place your legs against the legs of the chair from the outside and try to approximate the legs.

Internal-External Rotation of the Thigh

Dynamic exercise: Stand on one leg and, with some additional weight on your neck (sandbag, etc.) turn from side to side. (External rotation is practiced automatically.) Change legs regularly.

Static exercise: Stand with the legs closed tightly (toe against toe, heel against heel); try to rotate the foot inward and outward against the resistance.

Extension of the Knee Joint

Dynamic exercise: Squat jumps of all kinds, knee bends.

Static exercise: Sit in a chair and cross your legs at the ankle; try to raise the hindmost leg against the resistance of the anterior part of the leg. Change legs.

Flexion at the Knee Joint

Dynamic exercise: In the prone position, raise the feet against pressure by a colleague.

Static exercise: same as above.

Internal-External Rotation of the Leg

Dynamic exercise: Rotational squat jumps.

Static exercise: Sit on a chair with the legs together (inner edges are in contact), alternate pressure of the foot edges against each other.

Plantar Flexion in the Upper Ankle Joint

Dynamic exercise: Squat, high jump.

Static exercise: Sit on a chair and cross your legs so that the ball of one foot rests on the instep of the other. Push with dorsum of the lower foot against the ball of the upper foot. (In the upper leg, the extensors of the ankle are conditioned.)

Dorsal Extension in the Upper Ankle Joint

Dynamic exercise: Dorsal extension against a pull by a colleague (from the position of extension).

Static exercise: As above, with upper foot.

Pronation and Supination of the Foot in the Lower Ankle Joint

Dynamic exercise: Ski turn jumps—back and forth.

Static exercise: Sit on a chair, knees flexed, the balls of the feet facing and touching each other. Press with the outer edge of the foot against the floor; this strengthens the supinators.

For the pronators: Press with the ball of one foot against the inner edge of the other foot; now try to raise the inner edge of that foot against the resistance.

Bibliography

Becker W., Krahl H.: Die Tendopathien. Stuttgart, Thieme, 1978.

Benninghoff A., Goerttler K.: Lehrbuch der Anatomie des Menschen. Neu bearbeitet von *H. Ferner* und *J. Staubesand.* München, Berlin, Wien, Urban & Schwarzenberg, 1975.

Franke K.: Knorpelschäden am Kniegelenk durch Fehlbelastung und Trauma. Medizin und Sport 19 (1979). 1–6 (Heft 1/2).

Kahle W., Leonhardt H., Platzer W.: dtvAtlas der Anatomie. Band 1: Bewegungsapparat. Stuttgart, Thieme, 1975.

Lanz L. von neu bearbeitet von J. Lang., Wachsmuth W.: Praktische Anatomie, 1. Band, Teil 3 und 4. Berlin, Heidelberg, New York, Springer, 1972.

Leutert H.: Anatomie Berlin, Wien, Urban & Schwarzenberg, 1975.

Nemessuri M.: Funktionelle Sportanatomie, Berlin, Sportverlag, 1963.

Rauber A., Kopsch F.: Lehrbuch und Atlas der Anatomie des Menschen. Stuttgart, Thieme, 1964.

Rohen J.: Funktionelle Anatomie des Menschen. Stuttgart, New York, Schattauer, 1979.

Schiebler T.: Lehrbuch der gesamten Anatomie des Menschen. Berlin, Heidelberg, New York, Springer, 1977.

Sobotta J., Becher H.: Atlas der Anatomie des Menschen, Band 1. München, Berlin, Wien, Urban & Schwarzenberg, 1972.

Tittel K.: Beschreibende und funktionelle Anatomie des Menschen. Stuttgart, New York, Fischer, 1978.

Voss H., Herrlinger R.: Taschenbuch der Anatomie. Stuttgart, Fischer, 1975.

Weineck J.: Optimales Training. Erlangen, perimed Fachbuchgesellschaft, 1980.

Index